JUAN FREILE &
MURRAY COOPER

BIRDS OF
ECUADOR
AND THE
GALÁPAGOS ISLANDS

A PHOTOGRAPHIC GUIDE

H E L M

LONDON · OXFORD · NEW YORK · NEW DELHI · SYDNEY

HELM
Bloomsbury Publishing Plc
50 Bedford Square, London, WC1B 3DP, UK
29 Earlsfort Terrace, Dublin 2, Ireland

BLOOMSBURY, HELM and the Diana logo are trademarks
of Bloomsbury Publishing Plc

First published in the United Kingdom 2023

A catalogue record for this book is available from the British Library.

Library of Congress Cataloguing-in-Publication data has been applied for.

ISBN: PB: 978-1-4729-9337-3; ePub: 978-1-4729-9338-0;
ePDF: 978-1-4729-9339-7

2 4 6 8 10 9 7 5 3 1

Design by Julie Dando

Printed and bound in India by Replika Press Pvt. Ltd.

To find out more about our authors and books visit www.bloomsbury.com
and sign up for our newsletters.

CONTENTS

ACKNOWLEDGEMENTS

For continued support in recent years Juan Freile thanks: Xavier Amigo, William Arteaga, Jorge Bedoya, Alex Boas, Elisa Bonaccorso, Dušan Brinkhuizen, Héctor Cadena, Luis Calapi, Luis Carrasco, Juan C. Crespo, Renata Durães, the Fundación para la Conservación de los Andes Tropicales (FCAT) crew, Juan C. Figueroa, Jorge Flores, Rossy Gaibor, Paul Greenfield, Esteban Guevara, Ben Haase, Jordan Karubian, Niels Krabbe, José M. Loaiza, Yolanda Luna, Darwin Martínez, Teresa Mendoza, Andreina Morán, the Nature & Culture International crew, Lelis Navarrete, Andrea Nieto, Jonas Nilsson, Edison Ocaña, Scott Olmstead, Leo Ordóñez, Fernanda Patiño, Paolo Piedrahita, Alegría Plaza, Sandra Plúa and Conaves, Glenda Pozo, Rebeca Rivas, Luis Salagaje, Manuel Sánchez, Tatiana Santander, Wilmer Shiguango, Boris Tinoco, Paúl Tito, Carolina Tosta and Julie Watson. Fuller acknowledgements of the people and institutions who shared knowledge, field experiences, ideas, etc. are included in the Helm Field Guide *Birds of Ecuador* (Freile & Restall, 2018). Special thanks to Jenny Campbell for her editorial support and to Roger Ahlman and Dušan Brinkhuizen for providing their superb photographs. As always, thanks to the family Freile Ortiz, including Marimba, and to Daniela Pastrana and Valentina Zambrano.

Thanks must also go to the network of national parks and protected areas, plus all of the private reserves and lodges, for allowing the 'raw material' for this book to have a protected space in which to hop, jump, fly, eat and nest. Without these wild spaces this collection would be impossible!

Waved Albatross pair on Isla Española, Galápagos.

INTRODUCTION

Birding in Ecuador is a wonderful experience for both the keen birder and the novice. Being one of the most biodiverse countries on Earth, Ecuador has plenty to offer nature lovers across its varied wild areas, including the superb Galápagos archipelago. Access to most regions and ecosystems is reasonably good as the country is small, and has a good road network, fairly low-priced transportation, lodging and food, excellent birding and naturalist guides, large protected areas, and very good tourist infrastructure virtually throughout the country.

This photographic guide is intended for nature lovers who are not 'addicted' to birds. It does not cover the entire avifauna of Ecuador, which would be impossible in a book of this size given that more than 1,700 species have been recorded in the country. For those seeking a complete guide to its birds, we recommend the most recent guide published by Helm (Freile & Restall 2018), as well as other books such as *Birds and Mammals of the Galapagos* by Brinkhuizen & Nilsson (Lynx Edicions, 2020) and earlier works by McMullan & Navarrete (*Fieldbook of the Birds of Ecuador*, Ratty Ediciones, 2017) and Ridgely & Greenfield (*The Birds of Ecuador*, Cornell University Press, 2001). Those interested in general wildlife might need a copy of Vásquez's photographic guide (*Wildlife of Ecuador*, Princeton University Press, 2017).

We have selected 332 species that travellers might 'easily' encounter in different areas throughout Ecuador. Choosing these species was not a simple task, and some missing species could easily have been included. Unfortunately, space constraints compelled us to omit some 'must-see' species, including several of our favourite birds and many key species from a conservation standpoint, like the 42 Ecuadorian endemics (species that breed in no other country). A few examples of these endemics are listed in the next section.

The 332 species presented in this photographic guide cover a representative sample from all geographic regions, most ecosystems and as many families as possible, including charismatic and classic tropical birds like parrots, toucans, hummingbirds, tanagers and quetzals. In addition to a description of each species' appearance and behaviour, we include a brief 'where to see' section in the accounts. Again, space limits precluded us from providing full details, but online platforms like eBird (ebird.org) provide complementary resources for those requiring additional information.

We also include a section on the 'best' birding sites. Again, the selection is somewhat incomplete because of the wealth of birding hotspots in Ecuador. However, we are confident of providing a fair coverage of the most remarkable areas for birders and nature lovers alike. Each site has reference coordinates and elevation. A few logistical details and representative species are mentioned for each site. We also include a summary of bird conservation in Ecuador, aiming to raise readers' interest in this topic while travelling through the country.

BIRD CONSERVATION

Ecuador is full of natural wonders. Extensive Amazonian rainforest covers very large expanses of the east, whilst continuous, pristine Andean cloud forest shrouds steep slopes and deep creeks. Needless to say, the Galápagos Islands, including their coral reefs and other marine ecosystems, are among the best-preserved natural areas worldwide. Nonetheless, conservation problems are on the rise in nearly all regions of Ecuador.

Scarce natural habitats persist in most of the Pacific lowlands or temperate Andean valleys, while deforestation is spreading in the northern Amazon, the wet Chocó rainforests of Esmeraldas, and the Andean–Amazonian foothills. Main causes of habitat loss are intensive agriculture and cattle pastures, timber extraction, water pollution, slash-and-burn agriculture, and oil and mining exploitation, whereas unsustainable hunting and fishing are depleting species populations, and invasive species – predators, competitors, parasites, disease vectors – are causing serious declines in Galápagos birds.

Consequently, some regions are highly imperilled and in need of urgent conservation action. Timber extraction, mining, shrimp cultivation and oil palm agriculture threaten the Chocó region and adjacent mangroves. Intentional burning, large-scale agriculture (banana, rice, wood, maize) and cattle raising similarly affect the Tumbesian dry forests. Expansion of the agricultural frontier, mining and timber extraction put pressure on Andean cloud forests. Oil extraction, mining, timber extraction, oil-palm plantations and other large-scale agricultural practices threaten Amazonian rainforest. And, above all these specific problems, there is the gloomy shadow of climate change.

More than 180 globally threatened species occur in Ecuador; this figure includes two-thirds of the country's endemic birds and no fewer than 26 species that are currently on the brink of global extinction in the short term, being ranked as Critically Endangered or Endangered on the IUCN Red List. If we narrow the scope of analysis to those species at risk of extinction in Ecuador alone, the total rises to nearly 350 species. This means that one in five bird species that occurs in Ecuador might disappear from the country within a few decades. Although it is somewhat invidious to mention just a few cases, species with very restricted global distributions like Grey-headed Antbird, Scarlet-breasted Dacnis, Pale-headed Brushfinch, Grey-backed Hawk or Blue-throated Hillstar have tiny global populations. Meanwhile, more widespread species like Southern Pochard, Andean Ibis or Peruvian Thick-knee have tiny Ecuadorian populations that are in serious decline.

On the other side of the coin are the many conservation efforts currently underway; from mid- and long-term education and public-awareness campaigns, sustainable production at various scales of wide-ranging products such as coffee, chocolate, shrimp, tropical fruits, freshwater fish, seeds, grains, potatoes and other vegetables, maize, and many others on a wide array of protected land. Ecuador has a large national network of protected areas that currently comprises more than 60 national parks, reserves and recreational areas, covering roughly 20 per cent of the country's territory. This network includes enormous areas, such as Yasuni National Park (1 million hectares) and the Galápagos Marine Reserve (12 million hectares), to fragments that barely exceed 200 hectares. Although some regions in Ecuador, like the south-western dry forests, are insufficiently protected, others are covered by large and contiguous national parks, e.g. on the east Andean slope. Furthermore, private reserves are spread throughout Ecuador and currently cover several thousand hectares. Again, some private reserves are large and others tiny but located strategically where threatened biodiversity occurs. Given that many private reserves have tourist or capacity-building facilities, nature lovers have the opportunity to contribute to their management as visitors, volunteers or even sponsors. In the same vein, ancestral communities protect vast areas in the lowlands, Andean highlands and mangroves that are managed via their own policies and land-use practices. Nature tourism, including birding, can be a major contribution to this.

ENDEMISM

Ecuador has 42 endemic birds: 35 in the Galápagos and seven in its continental part. Endemism in the Galápagos includes seabirds (Waved Albatross, Flightless Cormorant, Galápagos Penguin, Galápagos Petrel, Galápagos Shearwater and Lava Gull), as well as unique landbirds such as Galápagos Hawk, Galápagos Martin, Galápagos Rail, Galápagos Dove, Lava Heron (sometimes ranked as a subspecies of the widespread Striated Heron), three flycatchers, four mockingbirds and 17 finches. Some of these are confined to a single island – the 'super-endemics' – like Vampire Ground Finch, restricted to two tiny islands in the far north of the archipelago, and two endemics on the small Genovesa Island. Remarkably, 23 of the Galápagos endemics are currently ranked as globally threatened; San Cristóbal Flycatcher is apparently extinct, and the total population of Mangrove Finch is just $c.100$ individuals.

Low endemism in continental Ecuador is explained by the small territory size. In fact, more than 200 species are mainly distributed in Ecuador, but their ranges just spill into adjacent Colombia and/or Peru. The seven endemics to Ecuador have minuscule ranges, the extreme cases being two hummingbirds: Black-breasted Puffleg and the recently discovered Blue-throated Hillstar. The remainder include two additional hummingbirds (Esmeraldas Woodstar and Violet-throated Metailtail), El Oro Parakeet, Ecuadorian Tapaculo and Pale-headed Brushfinch. All are globally threatened.

A note on the photos

All images show the adult form, unless stated otherwise. The key below explains the regularly used abbreviations:

Adult – ad.
Breeding – br.
Non-breeding – non-br.
Juvenile – juv.
Immature – imm.

MAP OF THE REGION

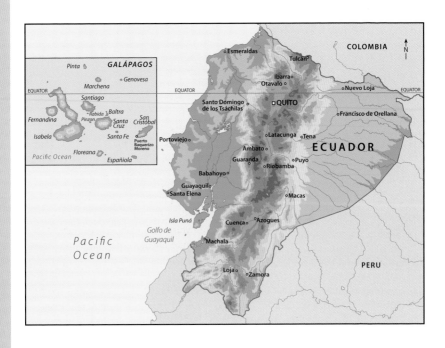

BEST BIRDING SITES IN THE REGION

Birding in Ecuador can be done nearly everywhere. However, some destinations are more suited to birdwatching due to their habitat quality and facilities for nature tourism, including bird-feeding stations. Here, we present summary information on some prominent birding sites for the experienced and new birder alike.

Chical–El Ángel
00°56′06″N, 78°11′09″W; 1,100–4,000m

Extensive patches of cloud forest connect with lovely *Espeletia* páramo in the north-westernmost Andes. Two roads traverse the area, one from Tulcán through El Angel Ecological Reserve, down to Maldonado and Chical, and another climbing from the town of Limonal, on the Ibarra–San Lorenzo highway, to the Gualpi plateau and then down to Chical. The best place in Ecuador for Purplish-mantled Tanager, Fulvous-dotted Treerunner and Beautiful Jay; Red-ruffed Fruitcrow occurs near Chilmá and Gualchán.

Espeletia *landscape at El Angel Ecological Reserve.*

Alto Tambo
00°54′51″N, 78°32′38″W; 600m

Six kilometres north-west of Alto Tambo, a mud-and-gravel road heads north. Good birding starts 1–2km further on.

Specialities include Chocó Woodpecker, Chocó Tapaculo, Yellow-green, Scarlet-and-white and Golden-chested Tanagers, and Black-tipped Cotinga. Mixed-species flocks can be overwhelming!

Playa de Oro
00°50′54″N, 78°46′56″W; 160m

The Playa de Oro community protects a vast expanse of wet Chocó rainforest that is connected to the extensive Cotacachi-Cayapas National Park. Tigrillo Lodge is 3km east of Playa de Oro and is only accessible by boat. A superb site for Chocó endemics; Berlepsch's Tinamou, Chocó Poorwill and Stub-tailed Antbird occur in the lodge's backyard. A good trail system provides opportunities to find Rufous-crowned Antpitta, the unique Sapayoa, Olive-backed Quail-dove, Five-coloured Barbet, Rufous-winged, Blue-whiskered, Emerald and Scarlet-and-white Tanagers, Plumbeous Forest-falcon, Plumbeous Hawk and Long-wattled Umbrellabird, among many other specialities.

Las Peñas–Yalare
01°06′44″N, 78°51′53″W; 15m

A roadside stop at Yalare, along the San Lorenzo–Esmeraldas highway, can be good for some species not regularly found in deeper rainforest. Stands of secondary forest are good for Cinnamon Woodpecker, Slaty-tailed Trogon, Blue Cotinga, Black-breasted and Pied Puffbirds, and Scarlet-breasted Dacnis. Some 46km by road south-west of Yalare is the beach town of Las Peñas. Nice mangroves near here are good for Mangrove Rail, Panama Flycatcher and Humboldt's Sapphire; the shrimp ponds are fantastic for migrant waders and waterfowl.

Andean cloud forest at the Intag Valley.

Intag Valley
00°20′54″N, 78°26′48″W; 1,300–3,400m

The Intag Valley lies in the Andes west of Cotacachi, off the beaten tourist path. Cloud forest remnants exist mostly on steep slopes and in remote locales. Intag is renowned for its strong local conservation and sustainability movement. Some good birding spots include Siempre Verde, Intag Cloud Forest, Los Cedros and Junín Reserves. Leymebamba Antpitta, Tanager Finch, White-rimmed Brushfinch and the endemic Black-breasted Puffleg (in the highest parts) are possible in this valley.

Yaguarcocha–San Pablo
00°22′17″N, 78°06′05″W; 2,200m;
00°12′31″N, 78°13′56″W; 2,700m

Two Andean freshwater lakes in Imbabura province, near Ibarra and Otavalo. Both are recreational areas for local people. Their shores have patchy reedbeds and mudflats; reedbeds at San Pablo are good for Subtropical Doradito and Virginia Rail, mudflats for boreal migrant waders. Waterfowl include local resident breeders like Slate-coloured Coot, Yellow-billed Pintail and Andean Teal.

Valle de Guayllabamba
00°03′32″N, 78°20′44″W; 2,100–2,300m

This warm valley 7km north of Quito's airport has a few patches of natural arid woodland that provide habitat for Purple-collared Woodstar, Giant Hummingbird, Buff-fronted Owl, Blue-and-yellow Tanager, Scrub Tanager and Streaked Saltator. Good spots for rather easy birding include Quito Zoo and Jerusalem Protected Forest. Creeks above Perucho are the only known locality for the possibly extinct Turquoise-throated Puffleg.

Pululahua–Calacalí
00°01′21″N, 78°30′08″W; 2,000–2,900m

The famous Mitad del Mundo tourist hub looks like poor habitat for birds, but the nearby town of Calacalí has some specialities, including the very rare White-tailed Shrike-tyrant, as well as Streak-backed Canastero and Band-tailed Sierra-finch along the Caspigasi trail. Pululahua Reserve is 6km north of Calacalí and has a few remnant patches of semi-humid forest, which are good for Maroon-chested Ground Dove, Leymebamba Antpitta, Buff-breasted Mountain-tanager, Rufous-chested Tanager and Buff-fronted Owl.

Nono–Yanacocha
00°06′42″S, 78°35′05″W; 2,700–3,600m

A hotspot near Quito, Yanacocha Reserve has long been known as the only site for the Ecuadorian endemic Black-breasted

Puffleg. This lush montane forest is superb for Andean rarities like Imperial Snipe, Undulated Antpitta, Andean Pygmy-owl and Crowned Chat-tyrant. Hummingbird feeders are visited by such specialities as Sword-billed Hummingbird, Sapphire-vented Puffleg and Buff-winged Starfrontlet, whilst Rainbow-bearded and Purple-backed Thornbills roam around. Zuroloma Reserve, further downslope, has worm feeders with Chestnut-naped Antpitta and Ocellated Tapaculo.

Tandayapa–Bellavista
00°05′57″S, 78°40′52″W; 1,700–2,300m
This renowned birding spot has a well-deserved reputation for its stunning cloud forest rich in endemic epiphytes. Watching birds along the road that traverses the area is fantastic, whilst lodges and reserves provide easy access to hummingbirds and other birds at feeders. White-faced Nunbird, Tanager Finch, Beautiful Jay, Flammulated Treehunter, Cloud-forest Pygmy-owl, Hoary Puffleg, Golden-headed and Crested Quetzals, Toucan Barbet, Plate-billed Mountain-toucan, Powerful Woodpecker, Semicollared Hawk, and up to 30 hummingbird species are a few specialities of the area.

Mindo
00°03′04″S, 78°46′42″W; 1,100–1,500m
One of Ecuador's main tourist destinations, Mindo provides plentiful experiences in adrenaline, nature, sustainability and birding tourism. Lodges, local operators and birding guides are top quality, and birding opportunities abound. It is also the perfect hub to visit other destinations like Bellavista, Milpe, Paz de las Aves and others. Easily seen birds include Sunbittern, Torrent Duck, Olivaceous Piha, Coopmans's Elaenia, tanager flocks

and many hummingbirds at feeders. Nearby spots include Sachatamia, San Tadeo, Santa Rosa, Séptimo Paraíso, Punto Ornitológico and Balcón Tumpiqui.

Paz de las Aves
00°01′11″N, 78°42′25″W; 1,700–1,900m
This small private reserve revolutionised birding by taming and feeding the elusive Giant, Moustached, Yellow-breasted, Chestnut-crowned and Ochre-breasted Antpittas, as well as Rufous-breasted Antthrush and Dark-backed Wood-quail. Other attractions include a hide in front of a bustling lek of Andean Cock-of-the-rocks, and fruit feeders for Toucan Barbet, Black-chinned and Blue-winged Mountain-tanagers, plus Black-capped, Golden, Beryl-spangled, Flame-faced and Golden-naped Tanagers. Hummingbirds include Velvet-purple Coronet and Violet-tailed Sylph.

Milpe–Pachijal
00°01′60″N, 78°52′06″W; 600–1,100m
At km 91 on the Calacalí–La Independencia highway lies a road to the Pachijal River. Forest patches and bird-oriented reserves along this road are good for foothill species of the Chocó region. Club-winged Manakin, Glistening-green Tanager, Crimson-bellied Woodpecker, Chocó Toucan, Blue-tailed Trogon, Purple Quail-dove and Black-tipped Cotinga are all regular, whilst the near-mythical Banded Ground-cuckoo is often seen in the area. Hummingbird and fruit feeders offer great birding opportunities.

Mashpi–Amagusa
00°09′35″N, 78°51′12″W; 520–1,300m
Extensive wet foothill forest provides excellent habitat for many Chocó endemics. The upper area, including Amagusa Reserve and Mashpi Lodge, is

accessed from Pacto town via a gravel road for 16km. Chocó Vireo, Indigo Flowerpiercer, Orange-breasted Fruiteater, Black and Rufous-brown Solitaires, Esmeraldas Antbird, Moss-backed and Glistening-green Tanagers, Golden-collared Honeycreeper and Baudó Guan are 'easy' targets. The lower area is accessed from km 104 along the Calacalí–La Independencia highway, by taking a side road for nearly 25km. Rufous-crowned Antpitta, Rose-faced Parrot and Long-wattled Umbrellabird are possible.

Silanche
00°08′43″N, 79°08′29″W; 360m
A gravel road starting at km 124 on the Calacalí–La Independencia highway reaches Río Silanche Reserve after 7km. Forest at the reserve is secondary and surrounded by cultivated fields, but provides good birding via its well-maintained trails, where Berlepsch's Tinamou is possible. Its canopy tower is well worth spending time at, as it is often surrounded by fast-moving flocks that include Rufous-winged, Golden-hooded, Blue-whiskered and Guira Tanagers, Brown-capped Tyrannulet, Slate-throated Gnatcatcher, Purple and Green Honeycreepers, and Orange-fronted Barbet.

Los Bancos–23 de Junio
00°01′55″S, 78°52′35″W; 1,150m
The small bustling city of Los Bancos has some small to mid-sized humid forest fragments around it and offers easy access to nearby birding hotspots like Milpe, Mindo, Mashpi and Silanche. One such spot is 23 de Junio, a small settlement 15km south of Los Bancos, that offers unique close-up views of displaying Long-wattled Umbrellabird. Ornate Hawk-eagle is also regularly seen here.

Quito and environs
00°11′11″S, 78°29′08″W; 2,700–3,800m
Birding in Quito might seem a bit unproductive, but several urban parks are actually good sites. The botanical garden is possibly the best urban spot, acting as a magnet for boreal migrants and local residents, including Blackburnian and Black-and-white Warblers, Masked and Black Flowerpiercers, Rose-breasted and Golden Grosbeaks. Climbing the Pichincha volcano via the Teleferico de Quito gives access to nice páramo with Black-billed Shrike-tyrant, Carunculated Caracara and Curve-billed Tinamou.

Canandé
00°31′34″N, 79°12′46″W; 300–500m
Located in south-central Esmeraldas province, Canandé protects 5,700 hectares of Chocó rainforest in an area highly threatened by logging and oil-palm agriculture. Habitat is similar, albeit a little 'drier', than Playa de Oro. Understorey flocks include such rarities as Ocellated, Bicoloured and Spotted Antbirds, and Banded Ground-cuckoo and Great Curassow can also be seen. Other Chocó endemics include Lita and Chocó Woodpeckers, Baudó Guan, Scarlet-breasted Dacnis and Blue-whiskered Tanager.

Mache–Chindul
00°22′23″N, 79°39′54″W; 200–800m
Some 30km west of the small, bustling city of Quinindé lies Mache–Chindul Ecological Reserve, which protects some large patches of southern Chocó rainforest. Two biological stations exist in the immediate vicinity, Bilsa and FCAT. The latter has great facilities for researchers and visitors, and provides easy access to beautiful rainforest, with Long-wattled Umbrellabird and Black-tipped Cotinga both common.

Lalo Loor–Cojimíes

00°05'16"S, 80°09'37"W; 0–350m

The north coast of Manabí province offers mangroves and estuaries near the coastal towns of Cojimíes and Pedernales, and an interesting mix of semi-deciduous and humid forest in the interior. Mangrove Rail and Rufous-necked Wood-rail are easy targets in the mangroves. Lalo Loor is a small forest reserve 27km by road south of Pedernales. Dry-forest species like Collared Antshrike and Ecuadorian Trogon occur in the lower part, and humid forest species occupy the higher portion; semi-humid sections harbour the spectacular Royal Flycatcher and Rufous-winged Antwren.

La Segua

00°42'11"S, 80°12'05"W; 5m

A large freshwater wetland next to a secondary road connecting Chone with San Vicente. Water level fluctuates throughout the year and the avifauna varies accordingly. Numbers of waterfowl, terns and waders can be impressive during boreal migration periods, and vagrants or very rare migrants often show up. Resident birds include Least and Pinnated Bitterns, Spotted Rail and the ubiquitous Black-necked Stilt, plus several egrets and herons.

Río Palenque

00°35'20"S, 79°21'48"W; 170m

An old field station where some of the first research on the birds of the Pacific lowlands was performed, but now seldom visited by birders and naturalists. Its 100-hectare forest patch is surrounded by plantations, but still proves good for some forest species, including Northern Barred Woodcreeper, Guayaquil Woodpecker and Orange-fronted Barbet.

Guayaquil and environs

02°06'18"S, 79°53'47"W; 0–300m

Birding in Ecuador's largest city can be exciting given its combination of dry forest, aquatic and mangrove species. Peruvian Pygmy-owl, Yellow-crowned Night-heron and Grey-cheeked Parakeet breed in the city. Its outskirts, including Cerro Blanco Protected Forest, Isla Santay or El Morro, offer fantastic birding experiences, Cerro Blanco being possibly Guayaquil's natural highlight. Red-lored Parrot, King Vulture, Blackish-headed Spinetail, Red-masked Parakeet, Henna-hooded Foliage-gleaner and Grey-backed Hawk are some of the species found there.

Los Frailes beach, in Machalilla National Park, with dry scrubland inland.

Machalilla–Ayampe

01°33'37"S, 80°48'16"W; 0–600m

The best-preserved dry forests by the sea occur in Machalilla National Park and immediate environs. The main hub is Puerto López, the departure point for Isla de la Plata, where Red-footed, Blue-footed and Nazca Boobies breed. Visits to Agua Blanca community, Los Frailes beach, Ayampe Valley and several other destinations and trails are also possible. The area has many spots for the endemic Esmeraldas Woodstar, and the birding is superb, with species like Saffron Siskin, Necklaced Spinetail, Ochraceous Attila,

Collared Warbling-finch, Ecuadorian Piculet and many others being easy to find.

Santa Elena Peninsula
02°13′23″S, 80°57′31″W; 0–300m

This arid peninsula, the westernmost point in Ecuador, combines xeric scrub with brackish wetlands and coasts. Peruvian Thick-knee, Sulphur-throated Finch and Burrowing Owl dwell in the scrub and open areas. Salt-evaporation ponds at Pacoa and Mar Bravo offer both numbers and diversity, with White-cheeked Pintail, Chilean Flamingo, Grey-hooded Gull, terns, phalaropes and many other waders present. Seawatching from La Chocolatera can be rewarding, as is a visit to Las Balsas community in the mountains inland. Ecuador's largest population of Red-lored Parrot and a few pairs of Great Green Macaw occur there.

Manglares–Churute
02°25′01″S, 79°38′58″W; 0–400m

Mangroves, dry forest and freshwater wetlands 56km south of Guayaquil along the Guayaquil–Machala highway. A good spot for Rufous-necked Wood-rail, Roseate Spoonbill and Horned Screamer in the mangrove and lagoons. Dry forest and scrub have their set of Tumbesian endemics including Rufous-headed Chachalaca, Jet Antbird, Elegant Crescentchest, Speckle-breasted Wren, Grey-and-gold Warbler and Black-capped Sparrow.

La Tembladera
03°29′20″S, 80°00′13″W; 15m

A little-explored wetland only 8km from Santa Rosa airport offers relaxed birding. Horned Screamer, Great Grebe, Masked Duck, Black-bellied Whistling-duck, Comb Duck, Wattled Jacana, White-cheeked Pintail, Cinnamon Teal, Snail Kite and Least Grebe are all possible, and numbers can be impressive during migration.

Zapotillo area
04°23′04″S, 80°14′27″W; 170–500m

Dry forest and scrub in extreme south-west Ecuador; the flora is simply marvellous, with kapok, golden goddess and *pretino* trees being highlights. The avifauna is less diverse than in more humid habitats, but many species are endemic to the Tumbesian region. Tumbes Tyrant, Tumbes Hummingbird, Slaty Becard, White-headed Brushfinch, Crimson-breasted Finch, Pale-browed Tinamou, Grey-backed Hawk, Ecuadorian Trogon, Pacific Parrotlet, White-tailed Jay, Scrub Nightjar and Plumbeous-backed Thrush are all reasonably common.

Seasonal dry forest during the rainy period.

Jorupe
04°22′17″S, 79°54′02″W; 540–800m

Jorupe Reserve protects 1,800 hectares of spectacular dry forest and is 5km east of Macará town. Surrounded by kapok trees, the lodge is simply stunning. A number of Tumbesian endemics are easy targets, including maize-fed Pale-browed Tinamou and Ochre-bellied Dove. Watkins's Antpitta, Henna-hooded Foliage-gleaner, White-tailed Jay, Peruvian Screech-owl and Blackish-headed Spinetail are all possible. Rice fields towards Macará are a good spot for Comb Duck.

Piñas–Buenaventura
03°39'14"S, 79°46'05"W; 750–1,100m

Piñas is located in the Andean foothills, surrounded by humid and semi-humid forests. Some 15km by road west of Piñas is Buenaventura Reserve, home of two Ecuadorian endemics: El Oro Parakeet and Ecuadorian Tapaculo. Additional species seen along the trails and access road include Ochraceous Attila, Grey-backed Hawk, Speckled and Rufous-throated Tanagers, Slaty-winged Foliage-gleaner and Black-winged Saltator. The hummingbird feeders can be simply overwhelming.

Cerro de Arcos–Saraguro
03°33'58"S, 79°27'23"W; 3,400–3,800m

An isolated cordillera west of Saraguro town is traversed by a road that connects Zaruma and Piñas to the west. Not on the birding circuit until 2018, when a new species of hummingbird was described: the spectacular Blue-throated Hillstar. A cottage near the top of Cerro de Arcos provides easy access to the hillstar's habitat. Mountain Caracara, Red-rumped Bush-tyrant and Blue-mantled Thornbill are also possible.

Utuana
04°22'01"S, 79°43'03"W; 2,500–2,600m

Some 16km by road east of Sozoranga, in extreme southern Ecuador, lies the small Utuana Reserve, which protects Andean semi-deciduous forest. A suite of species characterises Utuana, including Black-crested Tit-tyrant, Rainbow Starfrontlet, Purple-throated Sunangel, Bay-crowned and White-winged Brushfinches, Silvery Tanager and Black-cowled Saltator.

La Bonita road
00°28'15"N, 77°32'41"W; 1,500–3,000m

This road runs almost parallel to the border with Colombia in the eastern Andes, descending towards the Amazonian foothills. Good forest patches exist all the way, with side roads providing access to better-preserved tracts. Bicoloured Antpitta, Black-collared Jay, White-rimmed Brushfinch and Red-hooded Tanager are highlights of the upper part of the road. Handsome Flycatcher, Green-and-black and Black-chested Fruiteaters, and White-capped and Rufous-crested Tanagers are also possible. Mixed-species flocks can be spectacular.

Cerro Mongus area
00°26'10"N, 77°50'46"W; 3,200–3,600m

Montane forest on the east Andean slopes in Carchi province are continuous with those from the Colombian border south to Cerro Mongus. Access is slightly complicated and most likely to be attempted by the hardcore birder, but Nueva América to the south and Guandera to the north are easy to reach. Chestnut-bellied Cotinga, Masked Mountain-tanager and Crescent-faced Antpitta are the main targets, but Grey-breasted Mountain-toucan, Andean Siskin, Golden-crowned Tanager, Black-crested Hemispingus, Black-backed Bush-tanager and Black-thighed Puffleg are also possible.

Cuyabeno
00°00'43"S, 76°10'42"W; 200–250m

Cuyabeno Wildlife Reserve protects a large expanse of both terra firme and inundated Amazonian rainforest. Laguna Grande is the main attraction and is often visited during flooded periods. More than ten lodges are sited along the Cuyabeno River and at the lagoon itself.

Great boat-based birding but there are also some trails. Anhinga, Boat-billed Heron, Sungrebe, Orange-winged Parrot, Blue-and-yellow Macaw, Bare-necked Fruitcrow, Bat Falcon, Crested Owl and Blue-crowned Trogon are just a few examples of species to search for; the star, however, is the odd Hoatzin.

Limoncocha area
00°24′20″S, 76°37′09″W; 230m

Sited on the north bank of the Río Napo, 1.5 hours drive east of Coca, this biological reserve is surrounded by secondary forest and degraded land, but offers good birdwatching opportunities. Hoatzin, Greater Ani, Black-capped Donacobius, Limpkin, Snail Kite and Azure Gallinule are regular at the lagoon, whilst Gilded and Scarlet-crowned Barbets, Yellow-bellied Macaw and Black-fronted Nunbird are possible in nearby forest. Harpy Eagle has nested in the reserve, and White-lored Antpitta is often fed by a local guide.

Río Napo lodges
00°28′15″S, 76°27′33″W; 200–250m

The Río Napo provides access to some fantastic nature-oriented tourist destinations on both its north and south banks. Extensive forests still exist a few kilometres inland of the Napo. The best-known lodges are Sacha, La Selva, Sani and Añangu, but the full suite of choices is larger. Harpy and Crested Eagles are possible, as well as Collared Puffbird, Cocha Antshrike, Zigzag and Agami Herons, Blue-throated Piping-guan, Spangled and Purple-throated Cotingas, Golden-headed and Striolated Manakins, Opal-rumped and Opal-crowned Tanagers, and White-browed Purpletuft, to name just a few examples.

Yasuni–Shiripuno
00°40′28″S, 76°23′50″S; 200–300m

Yasuní National Park is the largest protected area in continental Ecuador: more than 1 million hectares. Some communities run ecotourism enterprises – notably Añangu – and two universities have research stations: Yasuní and Tiputini. Nearly 600 species have been recorded, including Black-necked Red-cotinga, Wire-tailed and Blue-backed Manakins, Grey-winged Trumpeter, Ringed Woodpecker, Wing-banded Antbird and more than 40 other antbirds, Salvin's Curassow, five macaw species, etc. Shiripuno River, south of Yasuní, is another notable ecotourism destination. Spotted Puffbird and Fiery Topaz are two prime examples of Shiripuno's rich avifauna.

Río Bigal
00°31′24″S, 77°25′12″W; 450–1,000m

This fairly recently created reserve on the east slope of Volcán Sumaco is accessed from Loreto. Its lodge is rather basic but adequate for nature lovers. A well-designed trail system and the access road offer superb birding; mixed-species flocks can be overwhelming. Bigal is possibly the best place in the world for Pink-throated Brilliant. Additional highlights include Pavonine Quetzal, Nocturnal Curassow, Fiery-throated Fruiteater, Amazonian Umbrellabird, Striated Antbird, Blackish Pewee, Large-headed Flatbill, Ecuadorian Piedtail, Chestnut-tipped and Golden-collared Toucanets, White-throated Woodpecker, Wing-banded Wren and many others.

Narupa
00°41′16″S, 77°44′10″W; 1,000–1,500m

Narupa Reserve is located in the lush Andean–Amazonian foothills at km 16 on the Narupa–Loreto highway.

A loop trail from the field station accesses Black Tinamou territories. Hummingbird feeders can be very productive, with Black-throated Brilliant, Golden-tailed Sapphire, Gould's Jewelfront, Many-spotted Hummingbird and Green-backed Hillstar being dominant species. Highlights along the access trail include Torrent Duck, Western Striolated-puffbird, Black-billed Treehunter and Sharp-tailed Streamcreeper. Orange-breasted Falcon often nests near the reserve entrance.

Volcán Sumaco
00°39′51″S, 77°35′04″W; 1,300–3,700m

This volcano, isolated from the main Andean cordillera, is almost completely covered in primary forest, with páramo at its top. Access is difficult to most areas, but there is one trail to the mountaintop – albeit a three-day hike. WildSumaco Lodge is adjacent to Sumaco Napo Galeras National Park. Shrike-like Cotinga, Black Tinamou, Jet Manakin, Foothill Screech-owl, Spotted Tanager, Yellow-throated Spadebill, Fiery-throated Fruiteater and Slaty-capped Shrike-vireo are a few notable species.

Cosanga–Guacamayos
00°35′24″S, 77°52′43″W; 1,800–2,300m

Las Caucheras road, 1km before Cosanga along the Papallacta–Tena highway, accesses San Isidro Lodge and Yanayacu Research Station. Roadside birding is great, but the trails and feeders in San Isidro are even better. Inca and Turquoise Jays, White-bellied Antpitta, Masked Trogon, Long-tailed Sylph, Rufous-banded Owl and Chestnut-breasted Coronet occur around the lodge. Guacamayos Pass, 8km south-east of Cosanga, is fantastic for the rare Peruvian Antpitta, Swallow-tailed Nightjar, Masked Saltator and Greater Scythebill. Several

nests of Black-and-chestnut Eagle have been found in the area.

Papallacta
00°20′37″S, 78°09′03″W; 3,100–4,100m

Papallacta lies just 73km east of Quito. It has several hot springs, but birders will be more interested in the road to Cayambe–Coca National Park, where Crescent-faced Antpitta, Masked and Black-chested Mountain-tanagers, and good mixed-species flocks are found. Some 10km below Papallacta is Guango Lodge, where the frantic hummingbird feeders are visited by Sword-billed Hummingbird, Glowing Puffleg and the furtive Mountain Avocetbill. At La Virgen Pass a road heads north to a high (and chilly) spot where Rufous-bellied Seedsnipe can be found.

The 5,700m-high Antisana volcano with surrounding páramo bogs and grassy fields.

Antisana
00°32′40″S, 78°13′32″W; 3,600–3,900m

An impressive volcano south-east of Quito and protected within Antisana Ecological Reserve. A paved road accesses a high plateau with Micacocha Lake and passes next to Tambo Cóndor, a nice lodge and restaurant in front of Peñon del Isco, where Andean Condor nests and roosts. Andean Ibis, Carunculated Caracara, Cinereous

Harrier, Andean Teal, Ruddy Duck, Andean Lapwing and Silvery Grebe are easily seen along the road.

Cotopaxi
00°36'47"S, 78°28'27"W; 3,700–4,000m

The iconic Cotopaxi volcano is one of Ecuador's main attractions. A road, departing at 75km along the Quito–Latacunga highway, accesses the base of Cotopaxi. Limpiopungo Lake offers good views of Andean Gull and the plateau is excellent for Carunculated Caracara, Black-chested Buzzard-eagle, Aplomado Falcon, Tawny Antpitta, Stout-billed Cinclodes, Ecuadorian Hillstar and Spot-billed Ground-tyrant.

Chimborazo
01°28'26"S, 78°50'30"W; 3,700–4,300m

Chimborazo, Ecuador's largest volcano, is surrounded by a dry plateau covered in arid páramo vegetation. Ecuadorian Hillstar is one of the main species to search for, along with Great Horned Owl, Black-winged Ground Dove, Rufous-bellied Seedsnipe, Slender-billed Miner and Plain-capped Ground-tyrant. A good road climbs Chimborazo, and several communities around the reserve run tourism initiatives.

Puyo area
01°28'46"S, 77°59'50"W; 900–1,300m

Puyo is 62km by road east of Baños, which is one of the main tourist hubs in Ecuador. Very good forest exists on the slopes above Mera, only 25 minutes west of Puyo, where several private reserves protect forests adjacent to Llanganates National Park. Grey-tailed Piha, Blue-rumped Manakin, Paradise Tanager, Wire-crested Thorntail, Coppery-chested Jacamar and Long-tailed Tyrant occur there. Tamandua

Reserve is another must in the Puyo area, with a fantastic canopy tower, bamboo-dominated forest and antbird-rich trails.

The mighty Pastaza at its upper section near Mera town.

Río Pastaza
02°27'46"S, 76°59'38"W; 250–280m

The poorly known lower Río Pastaza harbours extensive forest protected by local indigenous communities. Kapawi, near the Peruvian border, is the only lodge and is partially managed by local people. It offers fantastic birding opportunities, with rarities like Red-fan Parrot, Ancient Antwren, Orinoco Goose, Paradise Jacamar, Pavonine Quetzal and Grey-bellied Hawk.

Macas area
02°19'07"S, 78°07'02"W; 900–1,700m

Macas is in the Andean–Amazonian foothills, in the Upano Valley. The Cutucú cordillera lies to the east. This remote cordillera is home to Sharpbill, Rose-fronted Parakeet, Shrike-like Cotinga and many other major rarities. A road ascending the Andes to Guamote provides access to good forest birding, with great mixed-species flocks that can include Spectacled Prickletail, Blue-browed and Orange-eared Tanagers, Scarlet-breasted Fruiteater, Black-billed Mountain-toucan and many others.

Limón Indanza–Gualaquiza area

03°00′29″S, 78°32′21″W; 1,100–2,600m

Some 113km by road south of Macas lies the small town of Limón Indanza, and 90km further south is Gualaquiza. Both towns have roads that connect with Azuay province in the Andes. Birding along these roads can be very productive, with the transition from foothill to montane forests a must for birders and nature lovers. Black-and-chestnut Eagle, Cinnamon Screech-owl, Subtropical Pygmy-owl, Black-streaked Puffbird, Equatorial Greytail and Vermilion Tanager are a few examples.

Cuenca, Cajas and environs

02°52′01″S, 79°07′01″W; 1,900–3,600m

Cajas National Park, along the Cuenca–Puerto Inca highway, is the main natural attraction of Ecuador's third largest city. Hiking in this national park is awesome. Cajas is the best place in the world for the Ecuadorian endemic Violet-throated Metaltail. Other notable birds include Tit-like Dacnis and Giant Conebill. Andean forests at Mazán, Llaviucu, Camino al Cielo and other spots around Cuenca harbour populations of Red-faced Parrot and Grey-breasted Mountain-toucan. Yunguilla, 67km by road south of Cuenca, has the only remnant population of the Ecuadorian endemic Pale-headed Brushfinch.

Loja and environs

03°59′15″S, 79°11′54″W; 1,300–3,200m

The temperate valley of Loja is located midway between western dry forests and Amazonian humid forests to the east. Silvery Tanager, Koepcke's Screech-owl, Black-cowled Saltator and Bay-crowned Brushfinch breed within the city limits. The very dry Catamayo Valley to the west is good for Scarlet-fronted Parakeet, Drab Seedeater and Andean Tinamou. Cajanuma, 30 minutes south of the city, has very good flocks with Black-headed Hemispingus, Golden-crowned Tanager and Barred Fruiteater. Golden-plumed Parakeet occurs there.

Vilcabamba Valley

04°15′59″S, 79°13′24″W; 1,500–2,000m

This well-known tourist area, 42km by road south of Loja, is surrounded by dry woodland and scrub, with some notable Tumbesian endemics and other species like Plumbeous Rail, Peruvian Screech-owl, Chapman's Antshrike, Black-and-white Tanager, Andean Slaty Thrush and Three-banded Warbler. More humid forest patches near Vilcabamba are good for Bearded Guan.

Zamora area

04°05′27″S, 78°57′29″W; 900–1,000m

The southernmost Amazonian city in Ecuador is surrounded by beautiful foothill forest, with some extensive patches very close by. Spangled Coquette often visits city parks. The Bombuscaro area, ten minutes south of Zamora, is stunning. Amazonian Umbrellabird, Lanceolated Monklet and White-necked Parakeet are reasonably easy along the access trail. The nearby Copalinga Reserve has Grey Tinamou feeding on maize and a great show at the feeders, including Golden-eared, Green-and-gold and Golden Tanagers, and a bunch of hummingbirds.

Nangaritza Valley

04°22′09″S, 78°39′22″W; 900–1,800m

The landscape in this valley, 115km by road south-east of Zamora, is striking. Table-top mountains, part of the Cordillera del Condor, on one side, and Andean forested foothills to the west. This is a great location for Orange-throated Tanager, and also offers easy access to species confined to this cordillera, including Cinnamon-breasted Tody-tyrant, Roraiman Flycatcher, Royal Sunangel and Bar-winged Wood-wren. Additional rarities include Military Macaw, Sharpbill, Rufous-browed Tyrannulet, Solitary Eagle and Subtropical Pygmy-owl.

Tapichalaca

04°29'42"S, 79°07'56"W; 1,800–3,400m

The fantastic Tapichalaca Reserve lies at km 91 along the Loja–Zumba road and is home to the unique Jocotoco Antpitta, which visits feeders along with Chestnut-naped Antpitta. The reserve's trail system can offer good flocks, with Red-hooded Tanager, Masked Saltator, Black-throated Tody-tyrant and Orange-banded Flycatcher. The hummingbird community is large; a few examples include Little and Amethyst-throated Sunangels, Chestnut-breasted Coronet and Greenish Puffleg. Golden-plumed Parakeet and the superb Chestnut-crested Cotinga also occur.

Zumba area

04°52'03"S, 79°07'47"W; 800–1,300m

This remote area near the Peruvian border is largely deforested, but a few remaining patches have birds unique from an Ecuadorian perspective, many

Lava and white sand shores on Floreana Island, Galápagos.

of them endemic to the dry Marañón Valley that spans northern Peru: Marañón Thrush, Marañón Spinetail and Marañón Crescentchest. Tataupa Tinamou, Wedge-tailed Grass-finch, Buff-bellied and Black-faced Tanagers, the endemic race of Streaked Saltator, Northern Slaty-antshrike and Ocellated Crake are all reasonably easy to find.

Galápagos eastern islands: Española, Genovesa, San Cristóbal, Santa Cruz

Galápagos is simply matchless; one of the world's top nature destinations and a unique natural laboratory. The main hub for logistics is Puerto Ayora, on Santa Cruz. Many cruises and day trips depart from Santa Cruz, but the island itself offers plenty of birding experiences, including Galápagos Rail in the highlands. The eastern islands possess a range of island endemics including San Cristóbal Mockingbird, Genovesa Ground-finch, Genovesa Cactus-finch, Española Cactus-finch and Española Mockingbird. Waved Albatross breeds on Española.

Galápagos western islands: Fernandina, Isabela, Floreana, Santiago, Bartolomé, Rábida

The unique Galápagos Penguin and Flightless Cormorant are easy targets on Isabela and Fernandina, and the super rare Mangrove Finch occurs in tiny mangrove patches in both islands. Floreana has an endemic finch and mockingbird, which persist only on two tiny satellite islets. Galápagos Hawk, Galápagos Martin, Brujo Flycatcher and several finches occur in Isabela's beautiful highlands, whereas Greater Flamingo, the popular boobies, frigatebirds and other species are found along the islands' coasts.

Little Tinamou *Crypturellus soui* 21–24cm

Small, as its name implies, with greenish legs. Plumage varies from olive-brown to rufescent and dark brown; underparts are paler, more ochraceous or buffier; the crown is darker, and throat white. Not easy to see, but sometimes emerges into clearings and cultivated fields. When frightened, escapes rapidly on foot. Gives a clear, melancholic quavering song.

Where to see Widespread in the Pacific and Amazonian lowlands and foothills, mostly in humid secondary forest and dense shrubby areas nearby.

♂ east

Pale-browed Tinamou *Crypturellus transfasciatus* 25–27cm

The default tinamou of the Pacific dry forests, easily recognised by its pink legs and bold white eyebrow. Upperparts have bold blackish barring; underparts paler and plain. Gives an abrupt, piercing *oo-iiin!* throughout the day. Not easy to see, but emerges locally into cultivated fields near forest; also visits artificial feeders.

Where to see Dry forest and scrub in the south-west, sometimes where little forest remains; Jorupe is the best place.

♀

Torrent Duck *Merganetta armata* 38–42cm

Nothing similar in its unique habitat. Male is a beautiful duck with stylish black facial stripe and a streaky grey back; female all cinnamon below. Occupies clean-water, rushing rivers and larger streams in the east and west Andes, where family groups or pairs swim and dive in turbulent waters and rest atop boulders; only rarely escapes by flying.

Where to see Uncommon in forested rivers, mostly at 700–3,200m, along some of the birding routes descending from the Andes like Papallacta–Cosanga–Narupa, Tandayapa–Mindo, Atillo–Macas and Loja–Zamora.

Black-bellied Whistling-duck *Dendrocygna autumnalis* 46–53cm

Locally known as María, this duck is readily identified by its pink bill and legs, pale greyish face, cinnamon chest and black belly. In flight shows a bold white panel in the upperwing. Gregarious, sometimes in very large flocks together with Fulvous Whistling-duck. Feeds by dabbling, up-ending and grazing on shores. Flocks often move at night.

Where to see Common in the south-west lowlands, inland to several freshwater bodies in the central-west; more coastal in the north-west.

White-cheeked Pintail *Anas bahamensis* 44–50cm

The striking white face and red bill base of this beautiful duck are unique. In flight shows a green speculum; whitish on underside of wings. Mostly in small groups, but sometimes several dozen gather locally. Feeds by dabbling, head-dipping and up-ending.

Where to see Mainly coastal, in freshwater and shallow brackish wetlands; in Galápagos found in shallow bays, brackish lagoons and even freshwater ponds in the highlands.

Yellow-billed Pintail *Anas georgica* 48–57cm

Elegant duck of the highlands, with a characteristic yellow bill. Mostly warm brown, with heavily speckled underparts. In flight shows a blackish speculum bordered by two buffy stripes. Forms groups, sometimes large. Feeds by dabbling, up-ending, dipping head into water or even diving. Also grazes on grassy shores. Its population is apparently increasing.

Where to see Andean freshwater lakes and ponds, also artificial reservoirs like those at Quito airport and in nearby Cumbayá and Guangopolo valleys.

Rufous-headed Chachalaca *Ortalis erythroptera* 56–66cm

Chachalacas are noisy, as only chachalacas can be; their loud cacophony can go on and on. There are only two species in Ecuador, Speckled in Amazonia and Rufous-headed in the west. The latter has a rufous hood and rufous on the wings and tail. It moves in small groups at various heights, and often persists in small woodland patches.

Where to see Dry to semi-humid forests, woodland and forest borders in the Pacific lowlands and south-west foothills.

Blue-throated Piping-guan *Pipile cumanensis* 68–74cm

This handsome, large black guan is unique in Ecuador for its white crest and white patch on the wings. The whitish facial skin and bill, and blue dewlap, are also distinctive. Seen in pairs or small groups that hop and run along large branches in the forest canopy and borders.

Where to see Canopy of primary terra firme and flooded forests up to 400m elevation; the Río Napo lodges and Cuyabeno are good sites for it.

Sickle-winged Guan *Chamaepetes goudotii* 51–65cm

The blue face of this shy guan, combined with pink legs and rich rufous underparts, provide a unique suite of identification features. It makes a characteristic wing-whirring sound in flight. Moves in pairs or small groups at various heights above ground, but sometimes on the latter too. Attracted to fruit feeding stations.

Where to see Uncommon in cloud forest and borders in the west and east Andes; often seen along forested roads that transect the Andes.

west

fftfor tort in

Greater Flamingo *Phoenicopterus ruber* 105–120cm

Flamingos are classic 'exotic' birds. Greater occurs in Galápagos, where the total population is small. This lanky, long-legged flamingo is bright pinkish-red, has a tricoloured bill and pink legs. The smaller, duller Chilean Flamingo occurs locally on the south-west coast of Ecuador. Feeds with entire bill, sometimes even head, submerged, moving the bill sideways to filter plankton.

Where to see Brackish, shallow lagoons and ponds, also mangroves, on Floreana, Santa Cruz, Isabela, Santiago and Rábida islands.

Pied-billed Grebe *Podilymbus podiceps* 28–33cm

Four grebe species occur in Ecuador, one on high-Andean lakes and three in the lowlands; Pied-billed is the most abundant. It has a thick, whitish bill with a blackish ring near tip. Entirely grey, with blacker crown and throat. Single birds, pairs or small groups occur on freshwater bodies with abundant floating vegetation; excellent swimmer and diver.

Where to see Common in lakes, ponds and reservoirs in the Pacific lowlands, a few Andean lakes, and salty and freshwater Galápagos wetlands.

Pale-vented Pigeon *Patagioenas cayennensis* 28–31cm

The most widespread and easy-to-see pigeon in non-forested habitats. It is mainly vinaceous; the head and rump are ash-grey, belly whitish, and outer half of tail is pale. Gives a slow, mournful cooing from exposed perches. Often seen in small groups atop trees, especially where there is plentiful fruit.

Where to see Common in the Pacific and Amazonian lowlands, including flooded forests and forested lagoons and wetlands, like Cuyabeno, Limoncocha and Las Peñas.

Croaking Ground Dove *Columbina cruziana* 16–18cm

Small and distinctive with its bold yellow base to bill, black spots on wing-coverts, a maroon bar on scapulars and black flight feathers. Found in pairs or small groups in open, arid areas, including city parks and streets, beaches and barren terrain; can be confiding. Utters an odd buzzy croak.

Where to see Dry areas in south-west lowlands, southern arid valleys to 2,200m elevation, and north along the coast to the Colombian border.

White-throated Quail-dove *Zentrygon frenata* 29–32cm

This handsome quail-dove is more often heard than seen, but readily visits artificial feeders. It has a bold head pattern with a bluish-grey crown, buffy face, black moustachial, white throat and striped neck sides. Often seen alone, but also in loose pairs, preferring to walk away from observers; visits feeders locally.

Where to see Cloud forest in both Andean cordilleras up to 2,600m elevation; classic sites include Paz de las Aves, Bellavista and Tapichalaca.

Eared Dove *Zenaida auriculata* 25–26cm

One of the most abundant bird species in Ecuador, the ubiquitous Eared Dove has a distinctive black crescent on the cheeks and variable black spots on wings. Individuals vary from dull and dirty greyish to vivid ochraceous. Regular on wires, houses, agricultural fields, etc.; sometimes in large groups. Like all doves, takes off abruptly and noisily when alarmed.

Where to see All non-forested habitats in the dry Pacific lowlands and temperate Andean valleys.

Galápagos Dove *Zenaida galapagoensis* 18–23cm

The only dove in Galápagos is also very beautiful owing to its bright blue orbital skin, white cheeks, boldly streaked upperparts, vivid pinkish-brown breast and bright red legs. It is largely terrestrial, and amazingly tame like most Galápagos birds. Mostly found in pairs or small groups, sometimes alone.

Where to see Mainly in the arid lowlands, including coastal scrub and cacti-dominated woodland on all islands; rarer in the humid highlands.

Greater Ani *Crotophaga major* 46–48cm

This long-tailed, heavy-billed ani looks glossy in good light, combining purplish, greenish and blue; its eyes are pale yellow. Forms fairly large groups in the Amazon that move noisily and appear ungainly, often low above water. Several females attend a communal nest. Gives various growls and gurgles, some oddly electronic.

Where to see Mainly in flooded forest at lagoon and river edges in Amazonia; more locally in swampy areas of the Pacific lowlands.

Smooth-billed Ani
Crotophaga ani 33–36cm

Common and widespread in degraded areas, with a black bill that is compressed laterally, and the culmen arched and smooth. Moves in small to largish groups, active and nervous when approached; flight laboured, as if carrying a heavy load on its back. Often seen around cattle and horses.

Where to see Humid open areas in east and west lowlands to subtropics, to 1,800m; always spreads after deforestation. Replaced by Groove-billed Ani in the drier south-west.

Striped Cuckoo
Tapera naevia 28–30cm

This parasitic cuckoo has a rufescent erectile crest, long whitish eyebrow, dense streaking on upperparts, and long graduated tail. Quite shy when foraging low in bushes and grass, but uses exposed perch when giving its rising and rhythmic, easily imitated whistled song. Raises its crest and spreads its alula (a 'winglet' near the wing bend) nervously while vocalising.

Where to see Shrubby clearings, pastures and cropland in the Pacific lowlands and foothills; more locally in Amazonia.

Squirrel Cuckoo
Piaya cayana 40–46cm

An abundant cuckoo of forested areas, also ranges into cultivated fields and gardens. Uniform chestnut above, grey below; very long tail, with bold white feather tips. Yellow bill, orbital skin red in Amazonia, yellow in the west. Lively and bold; leaps among foliage with tail held loosely.

Where to see Common and widespread, up to 2,500m elevation; easy to see in Mindo, Tena and many other tourist hubs.

west

Band-winged Nightjar *Systellura longirostris* 20–23cm

Nightjars often perch along roadsides and trails; this is the default species in the Andes. Looks dark brown with buff spots on scapulars, white throat patch and narrow nuchal collar. A white band near tip of primaries and white tips to outer tail feathers (in males) are visible in flight. Often seen around lights.

Where to see Andean highlands up to 3,700m elevation, in forest borders, open areas and city parks.

Common Pauraque *Nyctidromus albicollis* 24–28cm

The commonest lowland nightjar is characterised by orange-chestnut cheeks. Upperparts grey to rufescent-brown, with black and buff spots on scapulars and wings. Males have extensive white in tail, females only white tips. Can remain still and allow close approach, but then flushes abruptly and flies in an erratic zigzag before landing again.

Where to see Common in lowlands to subtropics in east and west Ecuador, mostly in forest borders and adjacent clearings.

Oilbird *Steatornis caripensis* 43–48cm

The only nocturnal fruit-eating bird in the world. Heavy, all rufous with sparse white dots and 'diamonds'; heavy, hooked, pale bill. Inhabits caves and deep gorges, often in very large numbers, and groups produce a phantasmagorical cacophony of screams, snarls, clicks and growls; uses echolocation to enter and leave caves. Plucks oily fruit, like palms and wild avocados, in flight.

Where to see Mainly Andean slopes; well-known caves in Amazonian foot-hills near Macas, Sucúa and Archidona.

Common Potoo
Nyctibius griseus 34–40cm

As its name implies, the commonest potoo in Ecuador. Plumage varies from greyish to rufescent, always with dusky streaking, mottling, vermiculations and bars. Tail short, wing-tips almost reach tail tip. Perches motionless by day on stumps and fence posts, often low above ground. At night gives a mournful, haunting series of descending notes.

Where to see Widespread in lowlands, foothills and subtropics of west and east Ecuador, in forests and adjacent clearings.

brown morph

Great Potoo *Nyctibius grandis* 48–54cm

The largest of the potoos resembles a stub given its big-headed, chunky silhouette. It is greyish-white to brownish, all finely vermiculated dusky, and variably mottled and blotched. It perches alone (but sometimes in loose family groups) atop stumps or exposed branches, and remains motionless by day; at night, sallies from high perches and utters a far-carrying, grumpy scream.

Where to see Uncommon in the Amazonian lowlands, mostly in terra firme canopy and borders.

White-collared Swift *Streptoprocne zonaris* 20–22cm

The commonest and largest
swift is readily recognised by its
contrasting white collar, although
juveniles lack a complete collar.
Small to very large flocks, noisy,
often flying high overhead and
above many types of habitats,
including towns and cities; roosts
on cliffs near waterfalls. Flight is
fast and floppy. Outnumbers other
swifts in mixed-species feeding
aggregations.

Where to see Widespread from
lowlands to 4,200m elevation,
above wetlands, forest, open areas,
páramo, etc.

Fork-tailed Palm-swift *Tachornis squamata* 13cm

A small slim swift with a very long
and deeply forked tail that looks
pointed when closed. Underparts
dull whitish with duskier scaling on
sides and flanks. Forms rather small
flocks; flight fast and buzzy, with
shallow and very fast wingbeats
on stiff wings. Roosts and nests in
Mauritia palms.

Where to see Fairly common in
Amazonian lowlands and foothills,
mostly above forested areas, palm
stands and waterside clearings.

White-necked Jacobin *Florisuga mellivora* 9–10cm

A distinctive, stunning and common hummingbird; male has bright blue hood, white nape band and white belly to undertail. Female boldly spotted green on throat and chest, with white belly, scaly vent and broad white tail-tips. Conspicuous, active and pugnacious in defence of food resources, often dominating other hummingbirds at flowering trees and artificial feeders.

Where to see Common in humid tropics to lower subtropics (to 1,600m elevation) in well-known birding areas such as Mindo, Buenaventura or Sumaco; forest canopy and borders, secondary woodland, wooded gardens and hedgerows.

White-whiskered Hermit
Phaethornis yaruqui 12–13cm

Hermits are often much duller than other hummingbirds, but this species is exceptional due to its metallic green plumage. Its contrasting facial stripes, coral-red mandible and long graduated tail are diagnostic. Curious, like all hermits, it inspects observers at close range and escapes giving a sharp call. Non-territorial but rather bellicose.

Where to see Understorey of humid forest, borders and adjacent gardens in the western lowlands and foothills; regular at artificial feeders.

Great-billed Hermit
Phaethornis malaris 12–13cm

The commonest Amazonian hermit among six species has a red mandible, coppery-green upperparts, dull buff underparts, coppery rump and long graduated tail. Like White-whiskered Hermit, Great-billed is inquisitive and nervous. It feeds on *Heliconia*, bromeliads and other flowers with curved corollas, which it searches in a trap-lining fashion.

Where to see Typical hermit of forest understorey and borders in the Amazonian lowlands, mostly in terra firme forest up to 1,000m elevation.

Velvet-purple Coronet
Boissonneaua jardini 10–11cm

One of the most remarkable hummingbirds of Andean cloud forests; all glitter! Purple on forehead, throat and most underparts, and bluish-green upperparts. Underside of wings rufous; tail mainly white. Territorial and aggressive, often dominant at artificial feeders and rich nectar sources like *Inga* trees. Holds wings aloft after perching.

Where to see Regular at Santa Rosa, San Tadeo, Amagusa, Sachatamia and other well-known sites north-west of Quito.

Sparkling Violetear
Colibri coruscans 12–13cm

Ubiquitous hummingbird of Andean cities and towns, its endless metallic *tik-tik-tik-tik* song is part of the urban natural soundtrack. Entirely shining green; central belly and protractile ear-tufts violet-blue. Very aggressive and territorial against conspecifics and other species, including passerines. Performs a remarkable flight display, ascending high above exposed perches and swiftly diving down again.

Where to see Parks, gardens and cropland in Quito, Cuenca, Loja, Ibarra and all other Andean cities, towns and environs.

Black-throated Mango *Anthracothorax nigricollis* 11–12cm

Heavy-billed hummingbird. Male might look dark in the shade, but its shining blue throat sides and purplish-red undertail glow in good light. Female is more distinctive given the contrasting white underparts with a black central stripe. Amazonian and Pacific populations differ in tail pattern and the colour of its sides. Conspicuous and belligerent when 'guarding' flowering trees; often seen hawking insects at forest borders.

Where to see Uncommon in lowland habitats along large Amazonian rivers and lakes. In humid to dry forest and borders in the Pacific lowlands; regular in several coastal cities and towns.

♀

♂ east

Purple-crowned Fairy *Heliothryx barroti* 9–10cm

This lovely fairy is one of the few hummingbirds with all-white underparts, including the undertail. A black mask and purple crown (in males) are also distinctive. Only rarely visits artificial feeders, darts to large flowers, fans and lifts tail while hovering. Often robs nectar via holes in flower bases.

Where to see Forest borders, gardens and hedgerows in western lowlands and foothills. Typical sites include Playa de Oro, Mindo, Bucay and Buenaventura.

Purple-throated Sunangel *Heliangelus viola* 11–12cm

The striking male has a deep purple throat, bright blue forehead, bluish breast and deeply forked tail. Female has shorter tail, white throat with green spots; some have scattered purple feathers too. Bold and aggressive against other hummingbirds; chases and displaces them from rich flower patches or artificial feeders.

Where to see Fairly common in forest borders and shrubby clearings in the southern Andes, including woodland around Loja, Cuenca and other cities and towns.

Violet-tailed Sylph *Aglaiocercus coelestis* 10–11cm (excluding tail)

The stunning male has a very long tail (up to 12cm) that glitters from violet-blue to turquoise. Male in the north-west has a blue throat patch. Female's tail much shorter, belly cinnamon-buff, with a white band across chest and green spots on throat. This uncommon hummingbird occurs in cloud forests in the western Andes and is regular at artificial feeders, where dominated by more territorial species.

Where to see Regular in Tandayapa and Mindo regions, also Chical road and Buenaventura Reserve.

Black-tailed Trainbearer *Lesbia victoriae* 10–11cm (excluding tail)

An unbelievable little denizen of urban parks and gardens. The deeply forked tail surpasses 12cm in males but is shorter in females. Otherwise, this trainbearer has little decoration. The shining green throat patch of males can look dark in dim light. Highly territorial against conspecifics; males perform a spectacular aerial display in which the tail-streamers are exhibited and a ticking metallic sound is produced while shivering the tail feathers during deep aerial dives.

Where to see Common in Andean cities and towns up to 3,800m elevation.

Ecuadorian Hillstar *Oreotrochilus chimborazo* 11–12cm

One of the most spectacular birds of the high Andes. Males are characterised by a deep purple hood and immaculate white underparts. Female is dull green above and rather nondescript; underparts dirty buff-white with green dots on throat. Male of a subspecies restricted to dry páramos in central Ecuador has a distinctive green patch on throat. Largely confined to stands of *Chuquiraga*, with its fiery-orange spiky flowers.

Where to see Páramos from the Cajas region north to the Colombian border, above 3,600m.

♀

♂ central Andes

♂

Speckled Hummingbird *Adelomyia melanogenys* 8cm

A rather nondescript small hummingbird, identified by its dusky cheeks and bold whitish eyebrow. Underparts whitish with profuse green spots. Very energetic and belligerent despite its small size; feeds by hovering, but also clings from or perches next to flowers, and even robs nectar from holes made by flowerpiercers.

Where to see Abundant in montane forest, borders and adjacent clearings in the west and east Andes, sometimes even in tiny patches of habitat.

Rainbow-bearded Thornbill *Chalcostigma herrani* 10–11cm

Male has a nearly unbelievable, glittering, rainbow-like narrow beard. Its rufous crown does not glitter; long, steel blue tail has broad white tips. More discreet female has green-speckled throat and smaller white tail tips. Clings from or perches next to flowers with short corollas; its spike-like bill does not allow it to visit more tubular flowers; seldom at artificial feeders.

Where to see Montane forest borders, treeline and shrubby páramos in west and east Andes.

Tyrian Metaltail *Metallura tyrianthina* 8–9cm

A noisy little hummingbird readily identified by its coppery-red and slightly forked tail. Male metallic green, with more glittering throat. Female has buffy throat to breast, with sparse green spots. Highly energetic and 'brave'; despite its small size, fights larger species and chases them from territories. Feeds by hovering or clinging to flowers.

Where to see Common in forest borders, treeline and shrubland in the Andes. Outnumbers other metaltails.

Sapphire-vented Puffleg *Eriocnemis luciani* 11–12cm

The commonest and most widespread of the nine puffleg species in Ecuador. It has a long, deeply forked, blue-black tail, blue forehead and purple vent. Territorial but not very aggressive, it often tolerates other hummingbirds at rich nectar sources, including artificial feeders. Outnumbers other pufflegs in areas where three or four species co-exist.

Where to see Montane forest, borders, treeline and shrubby clearings in the Andes; fairly common at Cajas and Yanacocha.

Collared Inca *Coeligena torquata* 11–12cm

Very distinctive humming-bird despite the scarcity of 'glitter' in its plumage. Male velvet-black, female greener and shinier, both with a bold white chest patch; tail black and white. Bill distinctly long and straight. Boldly plumaged, even in dim light of forest interior; visits flowers with long corollas in forest interior and edges, also regular at feeders.

Where to see Fairly common in Andean subtropical to temperate forests, in prime birding localities like Bellavista, San Isidro, Chical, Cajanuma and Tapichalaca.

Shining Sunbeam *Aglaeactis cupripennis* 11–12cm

Inattentive observers might think this hummingbird boring because it lacks any glitter, but this perception changes when good light touches its lower back such that it glistens gold, purple, lime-green and yellow. A conspicuous and very territorial species that generally dominates other hummingbirds. Holds wings open long after perching.

Where to see Common in páramos, treeline and shrubby clearings in the Andes; good sites include El Ángel, Antisana, Cajas, Chimborazo and nearly all other páramos.

♂

Rainbow Starfrontlet *Coeligena iris* 11–12cm

As stunning as its name suggests, this species wears a unique rainbow on its crown. Its rich orangey underparts are also highly distinctive, but less extensive north of Azuay. Bill long and straight, like all starfrontlets and incas. Acrobatic and conspicuous in flight, visits a variety of flowers, mainly those with long corollas; also common at artificial feeders.

Where to see Common in temperate valleys of the southern Andes up to 3,300m; classic sites include Utuana and Saraguro.

♂ south

Great Sapphirewing _Pterophanes cyanopterus_ 15–17cm

Our second-largest hummingbird is characterised by its bright blue wings. Male all green; female has cinnamon underparts. Tail long and forked. Flight pattern distinctive: slow wingbeats and some glides. It briefly keeps wings open after perching. Feeds mostly by hovering at a variety of flowers; also visits artificial feeders.

Where to see Forest borders, treeline and shrubby páramos in the high Andes; classic sites include Yanacocha and Cajas.

♀

♂

Hummingbirds

Booted Racket-tail *Ocreatus underwoodii* 8–9cm

Small and oddly adorned, male of this species is unique among Ecuadorian hummingbirds for its tail rackets. Female lacks them, but distinguished by white underparts with heavy green spots; tail longish and outer feathers tipped white. West of the Andes has white leg tufts; tufts buffy in east Andes. Flight slow, wavy, with a characteristic buzzing sound, also cocks and fans the tail. Regular and combative at artificial feeders and abundant flowers.

Where to see Uncommon at many sites in foothills and subtropics in the west and east Andes.

♂ west

♀

Sword-billed Hummingbird *Ensifera ensifera* 13–14cm

The odd bill is almost as long as its whole body! Male all green; female has underparts with shining green spots. Feeds on very long tubular flowers at forest edges. Amazingly, it also manages to introduce its bill into the tiny hole of artificial feeders.

Where to see Rather rare in temperate zones (above 2,500m elevation) throughout the Andes, including well-known areas like Papallacta, Yanacocha, Guango and Pasochoa.

Fawn-breasted Brilliant *Heliodoxa rubinoides* 10–11cm

Brilliants are strong hummingbirds with long and heavy bills (for a hummingbird). This species is distinguished by its tawny underparts with a delicate pink throat patch (male). Female has green mottling in throat. Often congregates with other hummingbirds at rich flowering patches and artificial feeders, but is not as aggressive as other brilliants.

Where to see Forest understorey, borders and adjacent woodland in west and east Andean cloud forests like Bellavista or Cosanga.

Giant Hummingbird *Patagona gigas* 17–19cm

This well-named species is twice as large as many hummingbirds. Not colourful, but easily recognised by size, drab plumage, whitish rump and long tail. Its flight is strong, with slow wingbeats and short glides. Belligerent when *Agave* is flowering; also fond of terrestrial bromeliads and regularly seen at artificial feeders.

Where to see Temperate Andean valleys up to 3,300m elevation; regular around Quito, Ambato, Cuenca, Ibarra, and other cities and towns.

Short-tailed Woodstar *Myrmia micrura* 6cm

One of the tiniest hummingbirds in Ecuador, easily confused with a bumblebee. Breeding males have a violet-pink throat patch, white underparts and a white postocular stripe. Tail really is short. Female has whitish-buff underparts. Wasp-like flight. Not aggressive, approaching nectar sources discreetly and avoiding combat.

Where to see Common in xeric scrub, gardens, shrubby clearings and dry forest borders in the south-west lowlands; regular in Guayaquil and other coastal cities and towns.

Grey-breasted Sabrewing *Campylopterus largipennis* 11–12cm

A large and conspicuous hummingbird distinguished by its dull grey underparts and long forked tail with broad white tips. It can resemble the numerous Fork-tailed Woodnymph, but this sabrewing has a long bill and is notably larger. Solitary and not very aggressive; mainly in forest understorey, but might emerge into clearings and forest canopy.

Where to see Common around Amazonian lodges and reserves; locally to foothills, where often seen visiting artificial feeders.

Glittering-throated Emerald *Amazilia fimbriata* 8–9cm

One of the commonest hummingbirds in the Amazonian lowlands and foothills, especially in open areas. Red lower mandible distinctive, also white median stripe on bright green underparts. Sexes similar, but male's throat and breast have a bluer sheen, and female's spotted underparts are duller overall. Often congregates with other hummingbirds at rich nectar sources like *Inga* trees or artificial feeders, where can become territorial.

Where to see Regular in gardens and shrubby clearings at Río Napo lodges and Amazonian foothill sites including Narupa, Sumaco, Bombuscaro, Nangaritza and Puyo.

Fork-tailed Woodnymph *Thalurania furcata* 9–10cm

Possibly the commonest forest hummingbird in Amazonia. Male glistens emerald-green and lavender-blue, with a deeply forked, steel blue tail. Female slightly smaller; underparts pale grey and tail less forked, with narrow white tips to outermost tail feathers. Mostly solitary, but aggressively defends territory; might visit artificial feeders in foothills. There are no feeding stations in the lowlands, but it regularly visits gardens and forest edge.

Where to see Expected at any Amazonian destination, up to 1,200m and even higher in some areas.

Amazilia Hummingbird *Amazilia amazilia* 9–10cm

The rufous belly and white underparts are distinctive; variable green spotting on throat. In southern Andean valleys there is rather little rufous on the underparts. Energetic and territorial, always 'on the move'. Sometimes shows up in unexpected places like tiny gardens beside beaches.

Where to see Common in dry scrub, shrubland, forest, gardens and cropland in the Pacific lowlands; also warm Andean valleys in Azuay and Loja. Frequent in Guayaquil, Manta and other cities.

♂ lowlands

Rufous-tailed Hummingbird *Amazilia tzacatl* 9–10cm

One of the most abundant hummingbirds in western Ecuador. Glittering emerald with a rich rufous tail. Distinct, mostly red bill. Female has green spots on throat and is duller. An aggressive and hyperactive hummingbird, dominant at artificial feeders and flower patches; constant 'humming' while hovering.

Where to see Common in humid forest borders, gardens, shrubby clearings and hedgerows in the Pacific lowlands, foothills and subtropics; also, temperate Andean valleys.

Golden-tailed Sapphire *Chrysuronia oenone* 9cm

It is almost impossible to be more colourful than this sapphire. Male has a glittering purple-blue hood, turquoise and emerald-green body, and golden bronze tail. Female lacks hood, and underparts are whitish with heavy green spotting; tail similar to male's. Often dominates feeding stations at foothill sites, where numerous and fiercely territorial.

Where to see Fairly common in Amazonian forest borders and adjacent clearings mainly below 1,200m, including classic destinations like Bombuscaro, Sumaco, Narupa, Puyo and Bigal.

Hoatzin *Opisthocomus hoazin* 60–70cm

Unique and spectacular, almost dinosaur-like. Large and heavy, with a long crest, bare blue face and very long tail. Hoatzins are gregarious and wary, always found near water; they moan and grunt when approached, then hop and bound heavily, and escape with a short, laborious flight. Young drop to the water when threatened, but swim and climb back into bushes using their unique wing claws.

Where to see Fairly common throughout the Amazon, mostly up to 600m elevation.

Grey-cowled Wood-rail *Aramides cajaneus* 33–40cm

Wood-rails are often found in swampy areas and near water. Long yellowish bill and pink legs distinctive. Grey-cowled has a grey hood, cinnamon upper belly and olive-brown upperparts. Single birds or pairs occur along forested river and lagoon shores, often emerging at the water edge on muddy banks. Utters loud cackling notes.

Where to see Uncommon in riparian habitats throughout the Amazonian lowlands; easy to see at Cuyabeno and the Río Napo lodges.

Slate-coloured Coot *Fulica ardesiaca* 39–44cm

This chunky, slate-grey waterbird with a little white on the vent sides is unmistakeable. Variation in bill pattern includes all white, a white bill with yellow frontal shield, and white bill with red frontal shield. Juveniles are dusky brown, and often move around together. Bill laterally compressed, unlike the typically flat bill of ducks. Paddles noisily before taking off; swims on open water, but also near shore and vegetation; dives for brief periods.

Where to see Every freshwater Andean lake has coots, and the species is sometimes numerous, e.g. at San Pablo, Yaguarcocha or Micacocha.

red-shield form

yellow-shield form

Common Gallinule *Gallinula galeata* 30–36cm

This widespread gallinule has a bright red frontal shield and mostly red bill. Head and underparts blackish-grey, upperparts browner, and undertail-coverts mainly white. Immatures browner and duller, with no red on bill. Swims actively and vigorously in freshwater ponds and marshy lakes, also walks on floating vegetation, cocking tail constantly.

Where to see Common in various wetlands in the western lowlands and temperate Andean valleys up to 3,200m elevation.

Purple Gallinule *Porphyrio martinica* 27–32cm

A beautiful and widespread gallinule, characterised by its deep lavender head and underparts, green back and white undertail-coverts. Bill mostly bright red, frontal shield bluish. Juvenile has buffy head and underparts, and a blue shoulder patch. Vigorous swimmer, also walks and skulks in aquatic vegetation and adjacent scrub.

Where to see Common in freshwater marshes, ponds, vegetation-fringed lakes and stagnant rivers in the Pacific and Amazonian lowlands, and temperate Andean valleys.

Limpkin *Aramus guarauna* 60–70cm

The long-billed, long-necked and long-legged Limpkin looks like a large heron. It is all dusky brown with profuse white streaking. Often seen alone or in small groups in various marshy and freshwater wetlands, where it feeds largely on snails. Flies with legs held dangling. Gives a loud *ca-rra-ooo* that is far-carrying.

Where to see Fairly common in lowlands of east and west Ecuador, mostly up to 400m elevation, including wet cultivated fields.

Grey-winged Trumpeter *Psophia crepitans* 45–62cm

An unmistakeable tall forest bird, which looks oddly hunchbacked. Its glossy black, velvet-like plumage and silvery rear parts are diagnostic. Moves in small groups on floor of primary forest, searching for fruit and small animal prey. Highly territorial but wary and reluctant to fly; when approached, escapes swiftly on foot.

Where to see Rare throughout the Amazonian lowlands, but its resonant booming calls are often heard in deep forest around lodges and reserves, mainly below 700m elevation.

Black-necked Stilt *Himantopus mexicanus* 35–40cm

Unique, slim and very long black bill and coral red legs, and handsome pied plumage. Always in groups, sometimes large flocks, wading in fairly deep water but also favours muddy shores. Wary, giving strident calls when flushed that serve as an alarm for other waders. When threatened, flies in compact flocks, circles around, and lands somewhere nearby.

Where to see Common in marshes, salt ponds, mangrove edges, estuaries and other wet habitats along the entire coast, including most major Galápagos islands.

American Oystercatcher *Haematopus palliatus* 40–44cm

This attractive wader has a long coral-red bill, yellow eyes and short pink legs. Its entire hood is black, upperparts dark brown, and belly, rump and underwing-coverts white. Often in wary and noisy pairs or small groups, on rocky and sandy beaches, and tidal pools, where searches for hidden invertebrate prey.

Where to see Uncommon along continental coast and on most Galápagos islands, mainly at quiet beaches and estuaries.

mainland

Southern Lapwing *Vanellus chilensis* 31–38cm

The long 'hair-like' crest is characteristic. Black-and-white face and underparts. In floppy flight shows a distinctive black-and-white wing pattern, and broad black tail band. It is apparently spreading in the western lowlands and foothills and temperate Andean valleys. Pairs to small noisy flocks in open fields, not always near water.

Where to see Wet grassy areas, including football pitches, also sandbars and river islands in the Amazonian lowlands; more locally elsewhere.

Andean Lapwing *Vanellus resplendens* 33–36cm

A distinctive wader of the high Andes. It has a pale grey hood, darker mask, and bronzy-olive upperparts with a purplish-and-green gloss. Moves in loose pairs, but sometimes forms larger flocks in swampy areas and short grassy fields. Gives a loud, querulous *glee-glee-glee…* in floppy flight.

Where to see Fairly common in open páramo and adjacent fields; characteristic of frequently visited sites like Cotopaxi, Antisana, Cajas or Chimborazo.

Wattled Jacana *Jacana jacana* 20–23cm

Unmistakeable wader of
freshwater wetlands. Adults are
black and chestnut (darker in
west), with striking greenish-
yellow flight feathers shown when
wings held open or in flight.
Legs and toes are very long and
greenish. A red wattle at the sides
of the bill is distinctive. Juvenile
dull brown and whitish, lacks
wattles. Common at lakesides
where floating vegetation is
abundant, sometimes in very
small ponds and puddles. Flight
laboured, with legs held dangling;
gives loud cackles.

Where to see Common in
lowlands, easily seen even
from roadsides.

juv.

east

Whimbrel *Numenius phaeopus* 38–46cm

One of the commonest and most distinctive waders that arrives in Ecuador from its breeding grounds in North America, the Whimbrel has a long, downcurved bill, dark-striped head and profusely streaked, whitish underparts. Seen in small groups, pairs or single birds, sometimes with other waders. Picks and probes for invertebrate prey on tidal mudflats, brackish lagoons and sandy beaches.

Where to see Common along entire continental coast and all Galápagos islands, but is less numerous.

Sanderling *Calidris alba* 20–21cm

This thickset little wader has a stout black bill and black legs. Its upperparts are plain pearly grey, underparts pure white. In flight shows a bold white bar on wing. Forms small compact flocks that are often seen dashing back and forth with the waves, picking for invertebrate prey.

Where to see Common boreal migrant along the entire continental coast and most Galápagos islands, mainly on sandy beaches, but sometimes on tidal mudflats and brackish lagoons.

non-br.

Spotted Sandpiper *Actitis macularius* 18–20cm

The commonest boreal migrant wader. Its slim and elongated body is diagnostic, but the safest identification character is the constant bobbing of its rear parts. In non-breeding plumage has greyish upperparts and chest-sides, and white underparts. Breeders are variably spotted below. Seen mostly alone, always near water, walking along shore; flies low above water with fast fluttering wingbeats.

Where to see Easy to see 'everywhere' from rocky rivers to tidal mudflats.

Greater Yellowlegs *Tringa melanoleuca* 29–33cm

Two yellowlegs species visit Ecuador during the boreal winter and are often seen together. Bill in Greater is notably longer than lateral head length, also thicker at base and paler than Lesser Yellowlegs. Otherwise, both species are very similar. Yellowlegs wade in rather deep water, but also walk along muddy shores and tidal flats.

Where to see Common on coasts, but also inland at freshwater and brackish wetlands throughout the country.

Andean Gull *Chroicocephalus serranus* 46–48cm

The only year-round resident gull in the Andes. In breeding plumage, adults have a glossy black hood. Non-breeders have a small dark patch on head sides and a white crescent around eyes. Flies gracefully, often above recently ploughed fields. Sometimes congregates in small to mid-sized flocks. Can be distinguished from boreal migrants Laughing and Franklin's Gulls – which might visit Andean wetlands – by its paler grey upperparts.

Where to see Regular at frequently visited lakes in the Andean valleys and páramos, including Micacocha, San Pablo, Yaguarcocha, Cajas, Limpiopungo and others.

br.

non-br.

Grey-hooded Gull *Chroicocephalus cirrocephalus* 42–44cm

This elegant gull is commonest on the south-west coast. It has a crimson bill, eye-ring and legs, whitish eyes, pearl-grey hood and mid-grey upperparts. Non-breeders have only traces of grey on hood, but a dusky spot on cheeks. Seen mostly on beaches and at fishing ports, but also at brackish wetlands further inland.

Where to see Often seen on the Santa Elena Peninsula, also numerous on Guayas and El Oro coasts.

Galápagos Gull *Leucophaeus fuliginosus* 51–55cm

One of the rarest gulls on Earth; population estimated at 150–300 pairs is confined to the Galápagos. Uniformly sooty-grey, with darker hood, white crescents above and below eyes. Bill and legs reddish-black. Juvenile sooty-brown. Often seen alone, in pairs or small groups, rarely away from coasts.

Where to see Most islands, regular in mangroves, bays, sandy and lava beaches, harbours, fish markets and sea lion colonies.

Brown Noddy *Anous stolidus* 42–43cm

A very distinctive tern with an elongated body, long dark bill, short legs and long tail. Mostly dark brown, cap is silvery; has a white crescent below eyes. Juvenile has darker cap. Feeds by hovering and dipping, but also steals food from pelicans; often forms large feeding aggregations with other seabirds.

Where to see Regular on rocky cliffs on several Galápagos islands; also seen at sea between islands given its highly pelagic habits.

Galápagos

Sunbittern *Eurypyga helias* 42–48cm

Nothing like the Sunbittern occurs elsewhere in the American tropics. Its long slender body, long bill, neck and yellow-orange legs, striped head, and complex pattern on the open wings and tail are completely unique. Single birds walk beside forested streams and lakes, even leaping on boulders in rushing streams and rivers.

Where to see Uncommon in Amazonian lowlands to foothills, and in western foothills to subtropics. Classic sites include Mindo, Narupa and Bombuscaro.

west

Galápagos Penguin *Spheniscus mendiculus* 50–53cm

The only tropical penguin is endemic to Galápagos. One of the smallest penguins in the world, it is characterised by its pink lower mandible, narrow white chinstrap and mottled breast-band extending to flanks. Juveniles drabber, with a pale face. Forages in upwelling zones near coasts, often seen on rocky and lava beaches.

Where to see Breeds at a few sites on Isabela, Fernandina and Santiago islands, but also seen away from them, including on Bartolomé and Floreana.

juv., ad.

Waved Albatross *Phoebastria irrorata* 88–92cm; wingspan 230–240cm

Unmistakeable albatross that only breeds on two Ecuadorian islands. It has a long yellow bill, white hood, yellowish tinge to the crown and neck, and underparts finely waved dusky. Highly pelagic, sometimes near fishing boats. Flies with deep wingbeats interspersed by short glides.

Where to see Largest breeding colony on Española, Galápagos, provides fantastic views of display behaviours. Only a few pairs nest on La Plata, off mainland Ecuador. Often seen at sea between the various Galápagos islands.

Magnificent Frigatebird *Fregata magnificens* 97–107cm

Our largest seabird, with a very characteristic silhouette of long forked tail, long narrow wings and long bill. Breeding male has a large red gular pouch. Female has a broad white pectoral band, but a black throat. Juvenile has white head to belly. Great Frigatebird is fairly common in Galápagos; males are very similar but their backs are greener. Female has a white throat; juvenile an ochre wash to head. Acrobatic in air, chases seabirds, boats and even fishermen, trying to rob prey. Patrols high overhead.

Where to see Common along the entire Ecuadorian coastline.

Galápagos Petrel *Pterodroma phaeopygia* 41–43cm; wingspan 90–102cm

A commonly seen seabird endemic to the Galápagos Islands. Long narrow wings with black markings on the underside. Its forehead is white, cap black, rump dark. Black-and-grey upperparts, white underparts. Flight fast, bounding, with shallow flaps and long glides; it swoops above waves, then steeply glides downwards.

worn

Where to see Colonies confined to higher parts of Galápagos' largest islands, but it is commonly seen at sea between islands.

Nazca Booby *Sula granti* 79–87cm

A bright white booby with an orange bill, black mask, tail and flight feathers. Juvenile has dusky bill, blackish-brown hood and upperparts, and white nuchal collar. Highly pelagic and not gregarious, sometimes seen from shore near rocky areas and cliffs. Nests on open ground on rocky islands.

Where to see Widespread in Galápagos, with well-known nesting colonies on Española and Genovesa islands. Often seen during whale-watching trips off mainland Ecuador; breeds on La Plata Island.

Blue-footed Booby
Sula nebouxii 76–84cm

The commonest booby in Galápagos and mainland Ecuador, diagnosed by its bright blue legs. Head has dense dusky streaking, but nape and upper rump white. Juvenile has solid brown hood, grey legs. More coastal than Nazca Booby, sometimes several gather at feeding frenzies, where they perform spectacular plunge-dives. Nests in open areas.

Where to see Common throughout Galápagos, also along entire mainland coast; breeds on La Plata Island.

Anhinga
Anhinga anhinga 83–89cm

A large waterbird with spear-like bill and long slender neck. Male glossy black with silvery plumes on wings. Female has buffy-white head and breast. Single birds or pairs perch high, wings open to dry in sun. Capable of swimming partially submerged; darts after prey with its bill. Sometimes soars high, tail fanned and neck extended.

Where to see Uncommon on rivers and freshwater lagoons in Amazonian lowlands; rarer at freshwater lakes and rivers in western lowlands.

Neotropic Cormorant
Phalacrocorax brasilianus 64–69cm

This abundant and widespread
cormorant is mostly glossy
black. Its bill is greenish yellow.
Juvenile has dusky-brown
upperparts, whitish underparts.
Highly gregarious in large rivers,
freshwater lakes, ponds, estuaries,
artificial wetlands, coasts, shrimp
ponds, and even rushing rocky
streams. Often perches atop snags
and boulders to dry wings. Good
diver, and also strong flier.

Where to see Common in the
Amazonian and Pacific lowlands,
also coasts and more locally in
temperate Andean lakes.

non-br.

Brown Pelican
Pelecanus occidentalis 117–132cm

This familiar pelican outnumbers
its larger counterpart, the Peruvian
Pelican, which is confined to the
south-west coast of mainland
Ecuador. Breeders have a white,
yellowish and chestnut head; also
red decorations on the bill. Non-
breeders duller, as are juveniles.
Gregarious and highly aerial,
always seems to be on the move;
forms typical 'V'-shaped flocks
in flight. Plunge-dives for food,
but also submerges head while
swimming.

Where to see Common along entire
mainland coast and on all islands
of Galápagos.

br.

Herons and egrets

Rufescent Tiger-heron *Tigrisoma lineatum* 66–76cm

A beautiful heron of forested areas. Adult rufous chestnut with a narrow white median stripe on underparts. Juvenile very different, being rich buff with heavy blackish barring, white underparts with some barring on sides. At higher elevations Fasciated Tiger-heron occurs, which is very similar in juvenile plumage. Tiger-herons are solitary and rather lethargic, so easily overlooked unless flushed – often by a passing boat.

Where to see Regular, but difficult to see, in almost any forested river, lake and stream in the Amazon. Also at some freshwater wetlands in the Pacific lowlands.

juv.

Great Blue Heron *Ardea herodias* 100–130cm

Two forms occur in Ecuador, a rare boreal migrant and a Galápagos endemic subspecies. The latter is unique in the archipelago thanks to its black coronal plumes, greyish to buffy neck, bluish-grey upperparts and wings, and rufous thighs. Mainly seen alone at tidal ponds, rocky shores, brackish lagoons and mangroves, where patiently awaits prey.

Where to see Fairly common in coastal areas on most Galápagos islands; rarer and unpredictable in continental Ecuador.

Galápagos

Striated Heron
Butorides striata 35–43cm

Ubiquitous small heron characterised by a black crown, grey neck and breast-sides; upperparts greyish-green with darker streaks. The boreal migrant Green Heron differs only in green tone to upperparts and rich chestnut neck and breast-sides. Lava Heron, endemic to Galápagos, is almost blackish. Active hunter, crouches motionless to attack prey from perch.

Where to see Many kinds of freshwater wetlands, shady streams, coasts, mangroves and Amazonian rivers and lagoons.

Cattle Egret
Bubulcus ibis 46–52cm

The smallest white egret has a short, stout yellow bill; yellow legs become orangey in breeders, which also acquire a tawny tinge to crown, chest and back. This egret arrived in Ecuador less than a century ago, but has rapidly spread everywhere. Gregarious and highly mobile; regularly, but not always, around cattle. Often away from water.

Where to see Common throughout continental Ecuador, especially agricultural land in the lowlands and foothills, and Galápagos.

Snowy Egret
Egretta thula 53–68cm

This small and elegant white egret is readily identified by its slender black bill, yellow facial skin and black legs with yellow feet. The equally common Great Egret is obviously larger, with a heavier yellowish bill and dark feet. Active and gregarious, often chases prey or flushes prey by stirring water with its feet.

Where to see Common throughout mainland Ecuador up to 3,300m elevation, in freshwater and brackish wetlands; more coastal in Galápagos.

Yellow-crowned Night-heron *Nyctanassa violacea* 56–68cm

A common denizen of coasts in Galápagos and mainland Ecuador. The handsome bluish-grey adult has a boldly striped head pattern; the large orange eyes and heavy bill are also characteristic. Juvenile is brownish-grey with dense pale streaking on body and spotted wings. Juvenile Black-crowned Night-heron is browner and shorter-legged. Largely nocturnal, but active in day too. Feeds mostly alone but congregates at roosts. Gives a loud *kwaák!*

Where to see From mangroves and abandoned shrimp ponds to urban environments, including parks in Guayaquil and other large cities.

juv.

Roseate Spoonbill *Platalea ajaja* 71–80cm

A charming wading bird of coasts, tidal flats, mangroves and river edges. Its long spatulate bill, featherless head, pinkish legs and pink upperparts and belly are very distinctive. Breeders acquire a scarlet tone to the upperwings. Juvenile duller. Gregarious at feeding and roosting sites; moves bill sideways to filter-feed small planktonic prey. Elegant flight.

Where to see Mainly in mangroves, including those in Guayaquil and Esmeraldas cities; rarer on Amazonian rivers.

imm.

Andean Condor *Vultur gryphus*
100–130cm; wingspan 275–320cm

Majestic. Readily recognised by its bare pink head, white ruff collar and extensive white on upperwings. Juvenile brown, its head dusky and neck ruff pale brown. Mostly seen alone or in pairs, rarely in small groups, soaring high overhead. Roosts and nests on rocky cliffs.

Where to see The best site in Ecuador to see condors is Antisana, but the species is regular in other páramos and high valleys.

♂

Black Vulture *Coragyps atratus* 56–66cm; wingspan 130–160cm

Our smallest vulture is one of the commonest and most widespread birds. It has a bare blackish head and is entirely sooty black; in flight shows silvery windows in outer primaries. Tail short and wedge-shaped. Flies with fast wing flaps, and much soaring. Regular around garbage, fishing ports and roads, but also edges of large rivers.

Where to see Ubiquitous, easy to see throughout the country, mostly below 3,000m elevation.

Greater Yellow-headed Vulture *Cathartes melambrotus*
74–80cm; wingspan 165–180cm

A forest vulture characterised by its bare yellow head with a bluish crown, orange nape and black mask. Mostly sooty black, but flight feathers paler and greyer. The similar Lesser Yellow-headed Vulture occurs locally along the Río Napo, whereas the commoner Turkey Vulture is more widespread across the country. Soars low above the canopy, wings held in dihedral.

Where to see Uncommon above continuous forest and borders throughout the Amazonian lowlands and foothills.

Hook-billed Kite *Chondrohierax uncinatus* 38–44cm; wingspan 80–100cm

A very distinctive raptor that is apparently increasing in numbers. Its heavy hooked bill, white eyes and bicoloured orbital skin are diagnostic. Male plumage varies from ash grey to blackish. Females are browner above, neck and cheeks chestnut, underparts barred chestnut and white. Does not soar as long as other raptors, but glides low above ground on bowed wings. Feeds on snails.

Where to see Humid forest borders and adjacent cropland in east and west lowlands, including well-known sites like Tena, Mindo, Nangaritza and others.

♀ grey morph

♂ dark morph

Osprey *Pandion haliaetus* 54–59cm; wingspan 125–170cm

The most aquatic raptor is a migrant from North America. It has long, 'M'-shaped wings and a prominent head. Its dull brown mask and upperparts contrast with a white crown and underparts. Feeds on fish caught by plunge-diving; soars frequently, but also hovers powerfully and flies with deep wingbeats. Often perches atop trees, posts, stumps, etc.

Where to see Shallow bays, estuaries, freshwater wetlands, dams, large Amazonian rivers, mangroves; commoner in lowlands, rarer in Andean valleys and Galápagos.

Swallow-tailed Kite *Elanoides forficatus* 56–61cm; wingspan 120–135cm

Highly distinctive, elegant raptor, which resembles a large gull. Black-and-white plumage, long narrow wings and long, deeply forked tail. Flight graceful, with much gliding and soaring, often low above canopy. Sometimes gathers in dozens to hunt swarms of aerial insects; also descends to canopy to snatch prey.

Where to see Common above rain and cloud forests in west and east Ecuador, mainly up to 1,600m.

Snail Kite *Rostrhamus sociabilis* 40–45cm; wingspan 100–115cm

The archetypal raptor of freshwater wetlands, it feeds on aquatic snails picked from floating vegetation or the water surface. Bill thin but deeply hooked, with a bright orange base; legs orange. Male slate grey with white basal half of tail. Female and juvenile dull brown on upperparts, underparts whitish with dusky streaks, eyebrow whitish. Many congregate in some places, even at small ponds and creeks along roadsides and in cultivated fields.

Where to see Common in Pacific lowlands at places like La Segua, or Amazonian lakes like Limoncocha.

♂

juv.

♀♂

Ornate Hawk-eagle *Spizaetus ornatus* 58–68cm; wingspan 110–125cm

The 'commoner' of our large eagles has a long erectile crest and strong feathered legs; rufous cheeks and chest-sides, white throat, densely barred belly. In flight has broad wings and a banded tail. Often perches high in forest borders and adjacent clearings, but hunts inside forest. Soars long and high.

Where to see Well-known sites include 23 de Junio and Sumaco, but also seen in other forested areas in the Pacific and Amazonian lowlands and foothills.

west

Roadside Hawk *Rupornis magnirostris* 33–38cm; wingspan 65–90cm

Possibly the commonest raptor in Ecuador; easily recognised in flight by rufous windows near wing-tips. Its greyish head, upperparts and chest contrast with a barred belly. Familiarity with this hawk will help to identify rarer species. Seen on exposed perches even in midday heat. Frequently soars and glides.

Where to see This well-named species is easily observed at roadsides, clearings and river edges in east and west Ecuador up to 3,000m elevation.

Harris's Hawk *Parabuteo unicinctus* 48–54cm; wingspan 90–120cm

The lanky and long-legged Harris's Hawk is dark brown, with rufous shoulders and thighs, and white rump and tail base. Juvenile duller, heavily streaked below, paler rufous on shoulders, but tail pattern similar. Easily recognised in flight by tail pattern, rich rufous underwing-coverts and pale window near wing-tips. Regularly seen in pairs or trios above agricultural fields, scrub and towns; soars but also glides on cupped wings.

Where to see Andean valleys near Quito and other northern cities, also arid coastal lowlands.

juv.

Black-chested Buzzard-eagle *Geranoaetus melanoleucus*
60–69cm; wingspan 150–185cm

Massive, the largest true raptor in high Andes. Adult has slate-grey hood and upperparts, belly lighter and finely barred; wings broad, tail very short. Juvenile has longer tail, dark brown upperparts, paler and buffier underparts with dense dusky streaks. Masters Andean skies with long soars on flat wings and slow wingbeats. Often seen in pairs or trios above open fields and near rocky cliffs.

Where to see Fairly common in páramos and Andean valleys above 2,000m, in areas like Antisana, Pichincha, Cajas, Cotopaxi, Chimborazo and others.

juv.

Galápagos Hawk *Buteo galapagoensis* 45–56cm; wingspan 116–140cm

The only hawk in Galápagos, endemic and still rather common in more remote areas. Adult is dark brown, but shows whiter underside to tail and flight feathers. Juvenile has browner upperparts, and tawny underparts with bold dusky streaks. Found in various habitats from shores to humid highlands. Also versatile in food choice: carrion, vertebrates and large invertebrates.

Where to see Regular on Isabela, Fernandina, Santiago, Española and other islands; much rarer on Santa Cruz.

Burrowing Owl *Athene cunicularia* 20–26cm

The only terrestrial owl in Ecuador. Long feathered legs, rounded head, short tail and fierce expression. Varies from dark greyish brown to sandy brown above. Largely diurnal but also active at night. Often in pairs standing next to nest burrows; bobs up and down, and scrutinises observers when excited.

Where to see Arid open areas in south-west lowlands and temperate Andean valleys up to 3,000m; good places include Pululahua and Santa Elena Peninsula.

Barn Owl *Tyto alba* 36–40cm

The classic owl of popular lore and culture. Unmistakeable by overall shape, size, long feathered legs, heart-shaped face and brown eyes. Greyish to buffy upperparts, paler underparts with sparse brown spots. The Galápagos endemic form is notably darker above. Nocturnal in urban and semi-urban environments, but wilder and partially diurnal in Galápagos. An active hunter that searches for and pursues prey in low buoyant flight.

Where to see Historic towns with old churches and abandoned buildings in Pacific lowlands and Andes, up to 3,200m; rarer in Amazonia and Galápagos.

mainland

Galápagos

Owls

Peruvian Pygmy-owl *Glaucidium peruanum* 16–17cm

A lovely but fierce little owl with a large and distinctively rounded head. Has rufous, rufous-brown, brown and grey-brown morphs; all have white underparts with lateral streaks, but differ in amount of pale streaking and spotting on head, upperparts and wings. The only pygmy-owl in most of the Pacific lowlands; the similar Ferruginous Pygmy-owl replaces it in Amazonia. Diurnal and crepuscular, often seen perched on logs or stumps, being harassed by passerines.

Where to see Common in dry forests and open areas, including parks in cities and towns.

rufous morph

brown morph

Crested Owl
Lophostrix cristata 36–42cm

A spectacular large owl with prominent white ear-tufts, brown head and rufous cheeks. Underparts tawny, more ochraceous on belly. This strictly nocturnal and poorly known owl is often found on its day roost below palm fronds and in large-leaved trees. Its voice is a harsh, deep growl: *ggooOORRR*.

Where to see Humid forest, borders and adjacent clearings with tall trees in the Amazonian and Pacific lowlands and foothills. Amazon lodges sometimes have roosting pairs nearby.

east

Black-and-white Owl
Ciccaba nigrolineata 36–40cm

Clad in black and white, this large owl is very attractive; its yellow-orange bill also stands out. A similar species, Black-banded Owl, occurs in Amazonian lowlands and foothills. Strictly nocturnal, but often hunts moths and bats near lamps. Roosts in forest canopy and dense bamboo stands. Its varied repertoire includes a fast series of low *koo* notes.

Where to see Uncommon in humid to semi-deciduous forests and borders in Pacific lowlands and foothills. Well-known sites include Cerro Blanco, Buenaventura and Sachatamia.

Short-eared Owl *Asio flammeus* 36–40cm

Familiar owl of agricultural areas, with small ear-tufts and striking yellow eyes. Dark brown on upperparts, tawnier below, all coarsely streaked. The Galápagos endemic form is darker overall. Regularly perches on stumps or fence posts, then glides low above vegetation. In Galápagos found in a variety of habitats, including rocky seabird colonies.

Where to see Andean highlands and temperate valleys, including around Quito, Ibarra and Cuenca; also, most islands in Galápagos.

Galápagos

Ecuadorian Trogon *Trogon mesurus* 30–32cm

The white eyes of this trogon are distinctive. Males are bright green and red, with a black mask, white pectoral band and dark undertail. Females are grey and red, with a narrower pectoral band. Like all trogons, it forms loose pairs that remain quiet and stolid for long periods.

Where to see Semi-deciduous forest in the Pacific lowlands, to subtropics in south-west. Jorupe, Cerro Blanco, Ayampe and Lalo Loor are good sites.

♂

Golden-headed Quetzal *Pharomachrus auriceps* 33–35cm

Three species of quetzal occur in Ecuador; all are clad in emerald and red. Golden-headed and Crested broadly overlap in their geographic distributions, sometimes even foraging together at wild avocados and other fruit. Golden-headed differs mainly in its black undertail. Male has bright golden-green head. Female has dusky head and breast. Mostly in pairs that are easily overlooked if not singing or plucking fruit by laborious hovering.

Where to see Fairly common in foothills and subtropical forests and borders in east and west Andes. Classic species of cloud-forest roads near Tandayapa, Cosanga, Piñas and Macas.

Green-backed Trogon *Trogon viridis* 28–29cm

There are two trogon species in the Ecuadorian Amazon that have a yellow belly with blue-green (male) or grey (female) upperparts. Green-backed stands out for its pale bluish eye-ring and mostly white tail in males; females have black-and-white barring on tail. A similar species in the Pacific lowlands, White-tailed Trogon, was formerly considered a subspecies of Green-backed. Like other trogons, this species is lethargic and easily overlooked.

Where to see Amazonian lodges and reserves up to 1,200m, mostly in forest but also at borders.

Masked Trogon
Trogon personatus 25–26cm

Another green-and-red trogon in male plumage, with a white pectoral band, faint white barring on undertail, broad white tail-tips and a red orbital ring. Female is brown and red; the undertail is duller and orbital ring is white. Loose pairs, sometimes trios, perch calmly for long periods, with sudden 'attacks' on foliage for prey and fruit.

Where to see Forest and borders in subtropical and temperate regions in east and west Andes; Tandayapa and Tapichalaca are classic sites.

Whooping Motmot
Momotus subrufescens 36–42cm

Three motmots with blue diadems occur in Ecuador; Whooping is found only west of the Andes. Bright green upperparts, more ochraceous below. Its black mask and striking red eyes produce a peculiar expression. Single birds or loose pairs perch calmly in fairly open areas, moving their tail racquets sideways, then suddenly attack prey.

Where to see Lowland deciduous and semi-deciduous forest, hedgerows and woodland; to subtropics in south-west. Machalilla, Jorupe and Cerro Blanco are classic sites.

Rufous Motmot *Baryphthengus martii* 43–46cm

The loud resonant *hoop* calls of Rufous Motmot are key to identify it from the similar, but smaller Broad-billed Motmot, which has a nasal voice. Both are rich rufous, with a black mask, black pectoral spots and green upperparts. Green is restricted to the lower belly in Rufous Motmot. Sluggish like all motmots, it moves the tail sideways too; starts singing before dawn.

Where to see Humid forest and borders in Pacific and Amazonian lowlands and foothills.

Ringed Kingfisher *Megaceryle torquata* 38–41cm

The largest kingfisher is the only one to have bluish-grey and rufous plumage. Female differs from male in its bluish-grey and white pectoral bands; both have massive bills. Often seen perched high on exposed branches, wires or posts, flying high above ground, and plunge-diving for prey. Gives strident rattling calls in flight.

Where to see Common at a variety of freshwater wetlands in western and eastern lowlands and foothills, more locally in temperate Andean valleys.

Green Kingfisher *Chloroceryle americana* 18–20cm

The second-smallest kingfisher in Ecuador. Male has a rich rufous breast-band. Female has a narrow or broken green breast-band and green streaks on belly sides. Easily confused with the larger Amazon Kingfisher, which has a heavier bill. Other small kingfishers have more extensive rufous on the underparts. Inhabits forested rivers, streams and lakes, also more open ponds, artificial wetlands and mangroves, where it perches low and plunge-dives; flight fast, low and direct.

Where to see Fairly common and widespread in Amazonian and Pacific lowlands to foothills.

White-necked Puffbird *Notharchus hyperrhynchus* 25–26cm

The comical-looking puffbirds have large heads and appear neckless. This chunky species has a heavy bill, white front, throat, nuchal collar and belly, and is considerably larger than other black-and-white puffbirds. It perches motionless for long periods, inspecting its surroundings for prey, which is battered prior to ingestion.

Where to see Canopy and borders of humid forest in Pacific and Amazonian lowlands; often seen from canopy towers; edges of semi-deciduous forest in western Ecuador.

Barred Puffbird *Nystalus radiatus* 20–22cm

This beautiful puffbird is often heard before it is seen: a loud, rhythmic series of easily imitated whistles. It combines a rufous-chestnut crown and upperparts with a buff neck, nuchal collar and underparts; all are heavily barred. Often found in pairs perched on exposed branches or even wires; remains still for long periods, often moving its tail sideways.

Where to see Humid forest and borders in Pacific lowlands and foothills, in areas like Silanche or Canandé.

Black-fronted Nunbird *Monasa nigrifrons* 26–28cm

This bold, noisy nunbird of the Amazonian lowlands has a characteristic scarlet bill surrounded by puffy black feathers. It moves in groups of 4–6 birds, and is more active, vocal and nervous than puffbirds. Often follows monkey troops to capture flushed prey, but also joins mixed-species flocks.

Where to see Common in flooded and riparian forests, borders, adjacent clearings and river islands; easy to see along the Napo and Aguarico rivers.

Swallow-winged Puffbird *Chelidoptera tenebrosa* 16–17cm

This small puffbird looks like a very robust swallow. Its short bill, short neck and tail give it a peculiar silhouette. The white rump and underwings are clearly visible in flight. In small groups, but also single birds or pairs, atop exposed perches and performing long aerial sallies after flying insects, then usually returning to the same perch.

Where to see Easily seen at Amazonian river margins in the lowlands.

White-eared Jacamar *Galbalcyrhynchus leucotis* 19–20cm

Jacamars have long pointed bills, but only this species' bill is pink. The white ear patch contrasts with the chestnut body and blackish crown. A sluggish jacamar regularly found in forest borders; in some places several pairs occupy adjacent territories. Feeds on flying insects captured by rapid sallies and pursuits.

Where to see Amazonian lowlands, often near water; a regular sight by roadsides near towns and cities and along the Napo, Aguarico and Pastaza rivers.

Rufous-tailed Jacamar *Galbula ruficauda* 24cm

The common jacamar of the western lowlands, characterised by a large white (male) or buffy (female) patch on throat. Glittering coppery-green upperparts and breast, rich rufous belly and undertail. Loosely associated pairs sit on fairly exposed perches, including wires and posts; lethargic and calm as they watch for passing prey that is suddenly snatched in flight.

Where to see Humid forest and borders in the Pacific lowlands and foothills.

Gilded Barbet *Capito squamatus* 17–18cm

Strikingly plumaged like all barbets, with a rich orange throat, and yellow (male) or coarsely streaked (female) underparts; both have a yellow-olive crown and mainly black upperparts. Its resonant *joo-boop* notes are far-carrying and often ventriloquial. Mostly moves in pairs at canopy level, sometimes joins mixed-species flocks or congregates with other frugivores in fruiting trees; moves heavily through vegetation, often in outer foliage.

Where to see Fairly common throughout Amazonian lowlands and foothills up to 1,200m; regularly seen from canopy towers around Río Napo lodges.

Barbets

Red-headed Barbet *Eubucco bourcierii* 14–16cm

Very colourful even for a barbet. Male sports a scarlet hood, female a neat combination of orange, bluish and black on head; both have green back, rich yellow belly and heavy yellowish bill. Male in Amazonian subtropics has more extensive red in underparts. Active pairs often accompany mixed-species flocks in forest canopy and borders; their far-carrying, resonant *goo'oo'oo'oo'oo* (faster in Amazonian populations) is also distinctive.

Where to see Common in foothills to subtropics in east and west Andes, including Sumaco, Bellavista, Mindo, Chical and any road winding across Andean slopes.

Toucan Barbet *Semnornis ramphastinus* 20–23cm

A mix of eight different colours makes this species one of the most gorgeous birds in Ecuador. Its heavy pale bill has a black ring near tip. Mainly in pairs or small groups that clumsily hop along branches. Regularly with mixed-species flocks; fond of fruit, also visits feeders. Gives far-carrying, syncopated deep calls and bill-claps.

Where to see Cloud forest in the north-west Andes; well-known sites include Bellavista, Mindo, Chical, Intag and Amagusa.

Crimson-rumped Toucanet *Aulacorhynchus haematopygus* 35–37cm

The commonest of the green toucanets in Ecuador is the only one with red on rump. Bill dark red with a narrow white base. Mainly in pairs, which actively forage along branches, leaping up and down vigorously. Regular at fruit feeding stations, but also a merciless predator of small vertebrates.

Where to see Mostly in foothill and subtropical forest of the west Andes; classic sites include Mindo, Chical and Buenaventura.

Plate-billed Mountain-toucan *Andigena laminirostris* 42–44cm

Blue Andean toucans are simply stunning. A yellow plate near bill base is the most distinctive feature of this species, along with its bicoloured facial skin. Pairs or small groups occur in forest canopy and borders, clumsily leaping along branches in fruiting trees.

Locally visits fruit feeding stations. Gives loud nasal barking notes.

Where to see Confined to cloud forests in north-west Andes; well-known in Tandayapa, Intag and Chical regions.

Many-banded Araçari *Pteroglossus pluricinctus* 40–43cm

Araçaris are small slim toucans, of which four species occur in the Amazonian lowlands of Ecuador. As its name suggests, Many-banded is characterised by two black bands with red margins across its belly. It has a yellowish-and-black bill, greenish-blue facial skin and black head. Always found in flocks, which leap in an ungainly fashion along canopy branches.

Where to see Amazonian lowlands and foothills up to 800m in terra firme forest and borders.

White-throated Toucan *Ramphastos tucanus* 53–57cm

Toucans are quintessential tropical birds. The large White-throated differs little from the smaller and co-existing Channel-billed Toucan, and these species mirror the case of the yellow-bibbed species: the larger toucan barks, the smaller one croaks. White-throated moves in pairs or small parties in canopy, making clumsy leaps along branches and perching atop trees when vocalising.

Where to see Primary forest and borders in the Amazonian lowlands and foothills.

Black-mandibled Toucan *Ramphastos ambiguus* 52–56cm

This charismatic large toucan is characterised by a contrasting lemon-yellow bib and bicoloured bill. The smaller and co-existing Chocó Toucan is so similar that the safest way to identify them is by voice: Black-mandible barks, Chocó croaks. Pairs of Black-mandibled leap heavily in the canopy and vocalise from high perches.

Where to see Humid forest and borders in Pacific lowlands and foothills, also cloud forests on east side of the Andes.

west

Yellow-tufted Woodpecker *Melanerpes cruentatus* 18–19cm

This lovely little woodpecker has a unique head pattern with its yellow spectacles and red mid-crown. In flight, the white rump is distinctive. Abundant and widespread, it often forms groups but more regularly occurs in pairs. Active even in hot midday; also very vocal, regularly working in dead trees and snags. Fond of fruit like papaya.

Where to see Forest canopy, borders and adjacent open areas in Amazonian lowlands and foothills.

Black-cheeked Woodpecker *Melanerpes pucherani* 18–19cm

In a sense, Black-cheeked is the Pacific lowlands counterpart of Yellow-tufted. The head pattern of males is characteristic; note densely barred upperparts and belly. Female lacks yellow on forehead and red is limited to nape. Bold and widespread, on exposed branches and dead snags. Fond of papayas and other fruit.

Where to see Common in humid to semi-humid forest, borders and adjacent agricultural land in the Pacific lowlands and foothills.

Powerful Woodpecker *Campephilus pollens* 33–36cm

Largest Andean woodpecker. Black upperparts and chest interrupted by a white lateral stripe that meets in dorsal 'V'; rump white, belly chestnut with narrow black bands. Male has bright red crest. It has a loud nasal voice and its drum comprises three rapid, far-carrying knocks. Often seen in pairs foraging on trunks and large limbs, sometimes with mixed-species flocks. Nests in largish tree holes.

Where to see Andean forests and borders mostly below 3,000m, where uncommon. Good spots include San Isidro, Tapichalaca, Bellavista, Yanacocha and others.

Woodpeckers

Crimson-crested Woodpecker *Campephilus melanoleucos* 33–35cm

Male easily distinguished by its scarlet head, white fore-face, white 'V' over black back, and barred belly. Female has black crest and a broad white stripe on face. This large woodpecker moves in pairs, working large branches and trunks. Gives a fast, strident and nasal rattle, also drums 3–4 strokes in quick succession.

Where to see Amazonian lowlands and foothills, in primary forest and borders; the default large woodpecker along the Río Napo.

Crimson-mantled Woodpecker *Colaptes rivolii* 24–26cm

Few woodpeckers reach the highlands and temperate Andean valleys. This distinctive species has red upperparts, a yellowish face and yellow belly. Pairs or single birds work along trunks, snags, branches and limbs at various strata, sometimes with mixed-species flocks and regularly 'away' from proper forest.

Where to see Found in a variety of habitats along the Andes, including urban parks with abundant trees in Quito, Cuenca, Ibarra and other cities and towns.

Golden-olive Woodpecker *Colaptes rubiginosus* 21–23cm

A common and widespread woodpecker with a bold whitish face patch outlined by a red crown and moustachial stripe. Upperparts golden-olive; belly yellow with dense black barring. Female has a dark crown and moustachial stripe. Often seen in pairs, sometimes alone, and regularly with mixed-species flocks. Active, noisy, fond of papayas and other fruiting trees, but also behaves like a typical woodpecker, with much pecking and drumming.

Where to see Primary forest to borders and adjacent cropland along the entire coast and east slope of the Andes.

♀

♂

Black Caracara *Daptrius ater* 40–43cm; wingspan 90–100cm

Related to falcons, but not a prolific hunter like the latter. Readily recognised by its wholly glossy black plumage with little white near the tail base. Facial skin and legs reddish-orange, duller in juveniles. Flight appears strenuous, with continuous flapping and short glides; feeds on fruit, carrion and small prey. Gives irritating *krryaaaa* screams in flight.

Where to see Common in Amazonian lowlands to foothills, in forest and open areas alike.

Laughing Falcon *Herpetotheres cachinnans* 46–52cm; wingspan 75–90cm

A unique, large-headed and sturdy falcon with a characteristic dark mask contrasting with pale buff crown, collar, rump and underparts; upperparts brown, tail banded. Often perches sluggishly for long periods on exposed perches. Feeds mostly on snakes. Flies with shallow flaps and short glides above forest and open areas.

Where to see Common in humid to dry forest, woodland and adjacent fields in Amazonian and Pacific lowlands to foothills.

east

Carunculated Caracara *Phalcoboenus carunculatus*
51–56cm; wingspan 110–120cm

This elegant and charismatic, omnivorous raptor is readily identified by its black upperparts and dense black-and-white streaking below. Face red. In flight note pure white underwing-coverts. Sexes are similar. Juvenile dull brown, with yellow face and white panels near wing-tips (in flight). Soars erratically and flies fast, with strong wingbeats. Sometimes in small to largish groups on the ground, pecking soil, removing debris, making short pursuits after prey; also takes carrion.

Where to see Regular in Andean páramos above 3,000m from the Cajas plateau, near Cuenca, north to the Colombian border.

juv.

American Kestrel *Falco sparverius* 26–29cm; wingspan 50–60cm

The smallest falcon in Ecuador is characterised by its blue-grey crown, bold facial pattern, rufous back and tail; underparts buffy to ochre. Male has bluish-grey wings and plain underparts. Female has entire upperparts rufous with dark barring, underparts with dark streaks. Common in open areas, including towns and fields, where often seen hovering rapidly, also soaring and gliding, chasing bird prey in flight or attacking terrestrial prey from a perch.

Where to see Dry Andean valleys and coastal plains, but also deforested areas on Andean slopes.

Peregrine Falcon *Falco peregrinus* 38–48cm; wingspan 80–115cm

Mighty and imposing. Upperparts grey to slate-grey, underparts whitish variably barred blackish. Head pattern striking, with a broad blackish moustachial, and white face and throat in migrant subspecies, and a blackish hood in the resident subspecies. Flight powerful, with strong wingbeats, and spectacular swoops after doves, waders and other birds. Also glides and soars on slightly depressed wings.

Where to see The resident subspecies is confined to a few dry Andean valleys, including near Quito city. Two boreal migrant subspecies are regular on coast, but also in Andean valleys and Amazonian foothills.

resident

boreal migrant

Cobalt-winged Parakeet *Brotogeris cyanoptera* 19–21cm

A ubiquitous, small parakeet of deep forest and open areas alike. Looks chunky, with a short wedge-shaped tail; its cobalt-blue flight feathers are distinctive at close range, as are its orange chin and yellow forehead when perched. Flocks give strident calls in flight (*kliin-kliin-kilikilikili…*), sometimes high above ground. Often gathers with other psittacids at clay licks.

Where to see Possibly the commonest parakeet in Amazonian lowlands and foothills, even in cities and towns.

Rose-faced Parrot *Pyrilia pulchra* 21–23cm

Stunning, smallish parrot, with whitish bill, orbital ring and eyes. Its rosy face has a thin black outline. Head and chest coppery-brown. In flight shows a distinctive blue-and-red pattern on underwings. Mostly in small flocks, rather quiet while perched in fruiting trees. Twisting flight.

Where to see Uncommon in wet forests and borders of Pacific lowlands and foothills; easily seen in Playa de Oro, Amagusa and Mashpi.

Blue-headed Parrot *Pionus menstruus* 25–29cm

The typical parrot of the tropics; its bright blue hood is distinctive. Bill dusky, with a pink patch on upper mandible. Red vent unique among congeners. Forms small noisy flocks that regularly congregate with other parrots at clay licks. More discreet while foraging, but tends to use exposed perches for resting and calling.

Where to see Common, widespread and familiar in Amazonian and Pacific lowlands and foothills.

Bronze-winged Parrot
Pionus chalcopterus 27–29cm

Readily recognised because appears all dark. A closer look reveals its stunning bronzy wings, pale yellow bill, rosy-white chest patch, and purplish gloss to upper- and underparts. Moves mainly in small, noisy flocks, but several sometimes congregate at fruiting trees and even maize or plantain crops. Like other parrots, discreet while feeding but does sit on exposed perches at other times.

Where to see Humid and semi-deciduous forests, and adjacent clearings, throughout Pacific lowlands and foothills.

Orange-winged Amazon
Amazona amazonica 31–33cm

This parrot is regrettably a popular pet, making sightings in the wild even more memorable. Yellow-and-bluish facial pattern visible at close range; orange patch in flight feathers conspicuous, especially in flight. Forms small vocal flocks or moves in pairs. Its loud chattering when perched is remarkable.

Where to see Common in Amazonian lowlands, especially in flooded and riparian forests; flocks roost together on river islands.

Pacific Parrotlet *Forpus coelestis* 12–13cm

A tiny, charming and boisterous parrotlet, mainly lime-green. Males have a bluish-tinged nape and neck, deep blue in wings and rump. Females lack blue in rump and wings. Numerous in a variety of habitats from deciduous forest and woodland to beachside towns, urban parks, agricultural areas, etc. Fast-flying and noisy flocks number up to 50 birds or so.

Where to see Pacific lowlands and foothills, especially in dry south-west, even in areas with sparse vegetation.

Black-headed Parrot
Pionites melanocephalus 21–23cm

The semi-musical, ringing calls of this handsome parrot are highly distinctive. Its black crown contrasts with orange-yellow cheeks and throat, and darker orange nape. Belly white, thighs yellow, upperparts green. Forms small flocks that dart through forest canopy, sometimes emerging into clearings or above rivers. Rather tame when foraging; some flock members act as sentinels.

Where to see Fairly common in Amazonian lowlands, in terra firme and riparian forests, also river islands.

Red-bellied Macaw
Orthopsittaca manilatus 44–50cm

A small boisterous macaw, readily identified by its yellow facial skin, reddish patch on belly, and greenish-yellow underwings and undertail in flight. Moves in flocks wherein pairs often remain close together. Several individuals often gather atop palms, trees and decaying trunks. Feeds largely on palms, and even roosts and nests in them.

Where to see Flooded and riparian forests, and palm swamps, in Amazonian lowlands, sometimes near small towns and settlements.

Blue-and-yellow Macaw *Ara ararauna* 75–87cm

The commonest large macaw, and the only one clad in blue and yellow. Its raucous screams are far-carrying. Mainly in pairs, sometimes small flocks in which pairs remain close together. Flight fast and purposeful; often travels long distances between roosts and feeding areas. Often seen at clay licks together with Scarlet Macaw and smaller psittacids.

Where to see Forested areas in Amazonian lowlands; regular at Cuyabeno, Kapawi and Río Napo lodges.

Red-masked Parakeet
Psittacara erythrogenys 33–36cm

This charismatic parakeet of dry forests is unfortunately a popular cagebird, possession of which is illegal in Ecuador. A bright red mask and underwing patches contrast with the green body. Juvenile has little red on head. Mostly in flocks that sometimes congregate into larger groups. Noisy in flight, more silent when feeding, but individuals keep vocal contact when perched through constant raspy chatters.

Where to see The archetypal large parakeet of dry forests in the Pacific lowlands, including urban parks in Guayaquil, Portoviejo and other cities.

imm., ad.

Great Antshrike
Taraba major 19–20cm

Largest antbird in Ecuador, readily identified by its bright red eyes, stout bill and bicoloured pattern (male is black and white; female is rufescent and white). Reasonably easy to spot as it briefly emerges into open areas while foraging. Imitating its loud voice, an accelerating series of *coo* notes ending in a snarl, is often useful to lure it out of cover.

Where to see Humid and semi-humid forest borders throughout Amazonian and Pacific lowlands; good spots include Cerro Blanco, Machalilla and areas around Tena.

Streak-headed Antbird *Drymophila striaticeps* 14–15cm

Few antbirds reach the highlands. Of those that do, Streak-headed prefers bamboo stands. Male is white and rufous with dense blackish streaks on head, throat and breast. Female has a buffy, streaked head. Both have bold wing spots and long tails with broad white tips. Active and vocal pairs only briefly join mixed-species flocks; inspects bamboo tangles, dense foliage and dead hanging leaves.

Where to see Wet forest undergrowth and borders in east and north-west Andes; typical sites include San Isidro and Tapichalaca.

Lined Antshrike *Thamnophilus tenuepunctatus* 16–17cm

This elegant, barred antshrike is common in dense second growth, where not difficult to see if patient. In good light, its yellow eyes stand out from the black-and-white barred plumage, as does its stylish crest. Female is notably contrasting with rufous crown, wings and upperparts. Mostly in pairs in dense undergrowth, but regularly emerges into open; pairs maintain permanent vocal contact using loud nasal notes.

Where to see Common over entire Amazonian foothills and subtropics, up to 1,750m, including along roads near Tena, Archidona, Puyo, Macas and Zamora.

♂

♀

Collared Antshrike *Thamnophilus bernardi* 16–17cm

Dry forest and scrub are not ideal habitat for antbirds, but are home to this handsome species with a bushy crest. Male has black hood, white nuchal collar and boldly marked wings. Female has rufous crown and upperparts, and buffy underparts and nuchal collar. Pairs inhabit dense tangles and semi-open vegetation. Wags tail and raises crest constantly while roaming through the understorey, sometimes climbing higher.

Where to see Pacific lowlands, up to subtropics in the south. Classic species of arid scrub in Machalilla, Manglares-Churute, Jorupe and Zapotillo.

Pacific Antwren *Myrmotherula pacifica* 9–10cm

A tiny zebra-patterned bird of humid forest borders that enters gardens, hedgerows, cacao crops and the like. Difficult to say if black-and-white male is prettier than the orangey female; both share a streaky back and wings with two bold white bars. Moves in active pairs that glean invertebrate prey from foliage, underside of leaves, vine tangles and dead hanging vegetation. Its voice is a spritely series of inflected *chep* notes.

Where to see Fairly common in humid areas like Mindo, Silanche, Río Palenque, Playa de Oro, Santo Domingo, Bucay and others.

Chestnut-backed Antbird *Poliocrania exsul* 13–14cm

This common antbird gives a loud series of 2–3 ringing whistles. Bluish orbital skin. Male has head and most underparts slate-grey, chestnut back, wings duskier with small white dots. Female has rich orange-rufous underparts. Pairs inhabit overgrown vegetation and the ground; occasionally join flocks of army-ant followers.

Where to see Humid forest and woodland in Pacific lowlands and foothills. Silanche, Mache and Canandé are good sites.

White-plumed Antbird *Pithys albifrons* 12–13cm

A gorgeous, mostly chestnut antbird of deep Amazonian forests; its long white facial plumes are unique. An obligate ant-follower that travels through forest undergrowth with army-ant swarms; several individuals might gather at swarms, where they are restless and vocal. Clings to vertical saplings and darts to the ground after escaping prey.

Where to see Terra firme forest in Amazonian lowlands; often seen around Río Napo lodges and reserves, and other remote sites.

Common Scale-backed Antbird *Willisornis poecilinotus* 13cm

This noisy antbird of Amazonian forest undergrowth is characterised by heavy white scaling on the back. Male all grey with darker wings and tail, female similar but cinnamon-rufous. Mainly found in pairs; vocal and active, sometimes joins flocks of army-ant followers, but leaves flock as it abandons the species' territory; clings to vertical perches like White-plumed Antbird.

Where to see Inside terra firme forest in Amazonian lowlands and foothills.

Elegant Crescentchest *Melanopareia elegans* 14–15cm

Crescentchests are superb birds of dry scrub and woodland. This well-named species has a black head with long creamy eyebrow, creamy throat, black pectoral band, rufous edges to flight feathers and a long tail. Single birds or pairs in dense undergrowth, where not easy to see, but regularly in more open vegetation; frequently cocks tail.

Where to see Xeric coastal scrub and dry forest from south-west lowlands to subtropics. Regular at Machalilla, Zapotillo and Macará.

Giant Antpitta *Grallaria gigantea* 24–25cm

This spectacular antpitta is actually rare, but is included here because it is reasonably easy to see at a single locality two hours west of bustling Quito. It has a heavy bill, long legs and very short tail; olive and chestnut, with coarse wavy barring on underparts and a greyer crown. Largely restricted to wet forest interior; hops on the ground, always wary.

Where to see This and other antpittas are fed earthworms at Paz de las Aves.

west

Chestnut-crowned Antpitta *Grallaria ruficapilla* 19–20cm

Its loud, three-note song is commonly heard in forest and edges alike. Rufous-chestnut crown contrasts sharply with white underparts; breast-sides have long, broad dusky streaks outlined blackish. Like all antpittas, highly territorial. Pairs keep regular vocal contact while hopping through dense undergrowth; sometimes emerge into open areas.

Where to see Humid to semi-humid areas on both Andean slopes and locally in temperate valleys; also fed at Paz de las Aves.

Tawny Antpitta *Grallaria quitensis* 16–17cm

The only easy-to-see antpitta. Dull olive upperparts, tawny underparts, with paler orbital ring and loral area. Hops and bounds in open páramo, shrubby undergrowth and *Polylepis* woodland; perches on stumps, bunchgrasses or atop bushes to sing. Sometimes curious and confiding, attracted by whistled imitation of its distinctive voice.

Where to see Páramo grassland in the highlands above 3,000m, e.g. Antisana, Cajas, Pichincha, El Ángel, etc.

Ocellated Tapaculo *Acropternis orthonyx* 21–22cm

Not the commonest tapaculo, but the most spectacular by far. Blackish with white dots, a chestnut face, chest and rear parts; it has a longish tail, strong bill and very long claw on rear toe. Very shy but vocal; largely terrestrial, hops and bounds in dense undergrowth, scratching leaf-litter with both feet; very skulking, but sometimes crosses trails cautiously.

Where to see Wet montane forest undergrowth and bamboo stands; currently being fed at Zuroloma, near Quito.

Rufous-breasted Antthrush *Formicarius rufipectus* 18–19cm

Largely terrestrial. Chestnut crown, which is darker in the east, browner upperparts, black mask and rich rufous underparts; its short tail is often held cocked. Single birds or loose pairs walk, stop and walk again, bobbing their head regularly. Difficult to see; patient observers might get good sightings as it crosses trails slowly.

Where to see Wet forest and undergrowth in east and west Andes; fed at Paz de las Aves and eventually at artificial feeding stations elsewhere.

west

Plain-brown Woodcreeper *Dendrocincla fuliginosa* 20–21cm

Woodcreepers are dressed in shades of brown and are generally dull. This species lacks the streaked/spotted pattern of most woodcreepers, but has a distinctive dusky moustachial and pale face. Single birds or pairs often follow mixed-species flocks, but several gather at army-ant swarms, where vigorous and noisy. Tends to perch low.

Where to see Humid forest, borders and adjacent clearings in east and west lowlands and foothills; the default woodcreeper in many wooded habitats.

Wedge-billed Woodcreeper
Glyphorynchus spirurus 13–14cm

The smallest woodcreeper is also
the most abundant, but is sometimes
overlooked. Very small, with a short,
wedge-like bill. Narrow and short
buff eyebrow, plain buff chin, and
buffy spots on breast and sides. Single
birds or pairs rapidly hitch up vertical
branches and trunks in a rather
mechanical manner; often with mixed-
species flocks but sometimes away
from them.

Where to see Humid forest, borders
and adjacent clearings in Amazonian
and Pacific lowlands and foothills; easy
to see inside forest.

Long-billed Woodcreeper
Nasica longirostris 35–36cm

A fantastic and unmistakeable
woodcreeper with a very long, ivory
bill. Prominent white eyebrow and
bold white streaks on underparts.
Single birds or pairs easily attracted
by whistled imitations of their eerie
song. Probes into vegetation clumps,
bromeliads, bark crevices and
epiphytes. Often found in flooded
and riparian forests.

Where to see Amazonian lowlands,
commoner near water and in forest
with a fairly open canopy; Cuyabeno
and Limoncocha lagoons are
particularly good places.

Streak-headed Woodcreeper
Lepidocolaptes souleyetii 20–21cm

The default woodcreeper in forest borders and clearings in west Ecuador, readily recognised by its long, slender and decurved pale bill. Crown dusky with dense narrow streaks, and a narrow buff postocular stripe. Upperparts mostly plain rufescent brown; underparts heavily streaked buff; chin plain whitish buff. Conspicuous and confiding, regularly in pairs, which sometimes even forage and nest in human-made structures.

Where to see Western lowlands and foothills, including hedgerows, gardens and semi-urban areas.

Pale-legged Hornero
Furnarius leucopus 19–20cm

This familiar bird walks at a deliberate pace with characteristic head bobbing. Long pinkish legs, with long, slightly decurved bill, and pale eyes; greyish crown, long pale eyebrow, orange-rufous upperparts. Terrestrial and rather confiding, it has an explosive, strident song. Builds large mud-oven nests on trees and posts.

Where to see Widespread in open areas of the Pacific lowlands and foothills; in temperate Andean valleys and subtropics in Loja and Azuay provinces.

west

Woodcreepers and spinetails

Chestnut-winged Cinclodes *Cinclodes albidiventris* 17–18cm

Another familiar, terrestrial bird but this time found in the highlands. Has a short, near-straight bill. Brownish upperparts, rufous-chestnut patches on wings, and long white eyebrow; underparts greyish-buff, throat whiter and whitish spots on breast. Runs, hops and makes sudden stops, tail cocked; probes into soft ground, mud, vegetation or soil; rather tame.

Where to see Grassy and shrubby páramo, and adjacent agricultural fields, in Andes, up to 4,300m.

Pearled Treerunner *Margarornis squamiger* 15cm

This lovely species has a bold white eyebrow and pearly spots below. Back and wings rufescent brown; tail long, graduated and spiky. Single birds or pairs join mixed-species flocks; nimble and hyperactive, clambers and hitches along epiphyte-laden twigs and branches, inspects moss clumps, dead leaves and outermost foliage, sometimes upside-down.

Where to see Cloud forest, borders and treeline in the east and west Andes, e.g. at Cosanga, Yanacocha, Cajas and similar sites.

Many-striped Canastero *Asthenes flammulata* 16–17cm

As its name implies, this canastero is very streaky all over; has a buff postocular stripe and plain orange chin. Wings more rufescent, and tail long and spiny. Mostly terrestrial, walks and skulks in dense grass clumps, but climbs stumps and bush tops to sing.

Where to see Grassy and shrubby páramo in the Andes, up to 4,500m; often seen at Antisana, Pichincha, Cotopaxi and similar sites.

Red-faced Spinetail *Cranioleuca erythrops* 14cm

Arboreal spinetails are generally easier to see than their understorey counterparts. Its rich rufous head, wings and most tail feathers are distinctive; some have a faint pale eyebrow. Mostly in pairs, in high strata with mixed-species flocks; restless and energetic, searches and probes moss-laden branches and dead leaves, often hangs upside-down.

Where to see Cloud forest and borders in the west Andes and high coastal ridges; classic Mindo bird.

Necklaced Spinetail *Synallaxis stictothorax* 12–13cm

Very distinctive spinetail with long white eyebrow, white underparts, but narrow dusky streaks on breast; long, orange-rufous tail. More arboreal than its congeners, but also skulks in dense undergrowth; mostly in pairs, vigorously exploring branches, twigs, clumps and vine tangles. Very vocal, gives a loud and squeaky sputtering series.

Where to see Deciduous forest and arid scrub in Pacific lowlands, below 200m; regular in coastal scrub at Machalilla and Santa Elena.

south-west

Azara's Spinetail *Synallaxis azarae* 17–18cm

This common spinetail has a rufous crown, wings and long tail, a faint pale eyebrow, white moustachial and blackish scales on throat. Not easy to see as it skulks in dense thickets and undergrowth, but patient observers can get good views. Its 2–3 loud, squeaky notes are characteristic.

Where to see Woodland borders, shrubby clearings, regenerating scrub in the Andean cordilleras and temperate Andean valleys. Common around Ibarra, Quito, Cuenca and other cities.

White-bearded Manakin *Manacus manacus* 10–11cm

The handsome black-and-white male has no match in bird-rich Ecuador. Female and young male olive green, but their bright orange-red legs, like those of males, stand out even in the shade of the understorey. This manakin is unique among Ecuadorian species in the variety of fire-cracking mechanical sounds produced during collective male displays, where each male tries to perform best to attract visiting females.

Where to see Common in secondary forest and borders in the Amazonian and Pacific lowlands and foothills, where patient observers can acquire excellent views.

Wire-tailed Manakin
Pipra filicauda 11–12cm

♂

Lovely forest-dweller; even the drab olive-green female is unmistakeable due to her white eyes and long tail wires. Male's red, black and yellow plumage is adorned by longer tail wires. Males perform complex displays, including flips and about-turns, in which the tail wires are flaunted in the female's face, accompanied by some mechanical and vocal sounds. Both sexes are fond of melastomes and other small berries.

♀

Where to see Fairly common in deep forest throughout the Amazon, including lodges along the Río Napo, as well as Tiputini, Shiripuno, Cuyabeno and Kapawi.

Golden-headed Manakin
Ceratopipra erythrocephala 9–10cm

♂

Male's bright golden hood, reddish thighs and white eyes are prominent in the dim understorey light. Female dull olive, with pale bill and legs. Males perform courtship dances when females visit their display arenas, but are not in visual contact with other males.

Where to see Fairly common throughout Amazonian lowlands and foothills; known leks at Tiputini, Río Bigal, Limoncocha, Lumbaqui and Río Napo lodges.

Club-winged Manakin *Machaeropterus deliciosus* 9–10cm

An eccentric even among manakins, a family full of oddballs. Male has modified inner flight feathers utilised to produce a ringing, electronic sound by vibrating the feathers of one wing against the others. Plumagewise, it combines chestnut, red, black, white and yellow. The drab female has a distinctive buff-tinged face and yellowish belly. Fond of melastomes; its courtship display includes dances and the mechanical sounds already described.

Where to see Not difficult to see in foothills to subtropics, mainly in north-west; well-known sites include Milpe, Mashpi–Amagusa, Buenaventura and Chical.

Green-and-black Fruiteater *Pipreola riefferii* 18–19cm

This chunky cotinga is the commonest fruiteater in Ecuador and is stunning like all its congeners. Male has black hood, narrow yellow collar, green speckling/streaking on sides of yellow belly, and white edges to inner flight feathers. Female has green hood, yellow belly coarsely speckled/streaked green. Often in pairs or small groups, regularly with mixed-species flocks or with other frugivores at berry-clad trees. Rather sluggish, it utters a lovely piercing trill.

Where to see Fairly common in Andean cloud forests up to 3,200m, also forest borders. Good places include Bellavista and San Isidro.

Barred Fruiteater *Pipreola arcuata* 22–23cm

The largest of the plump fruiteaters. Male exquisitely combines a black hood with heavily barred underparts and yellowish decorations on its wings. Female lacks black hood; both have pale or red eyes, and red bill and legs. Rather lethargic, makes sudden flights to pluck fruit. May join mixed-species flocks.

Where to see Cloud forest and borders up to 3,300m in north-west and east Andes; good places include Yanacocha and Cajanuma.

Red-crested Cotinga *Ampelion rubrocristatus* 21–23cm

Elegant and attractive. Pale bill with a dark tip. Mostly slate-grey, head and wings darker; rump has whitish streaks. A long maroon-coloured, erectile crest decorates its crown. Often seen perched atop bushes and trees, regularly in pairs, erecting crest in displays; fond of mistletoes and small berries.

Where to see Humid forest, woodland, scrub and borders in Andes, including upper slopes of temperate valleys; classic sites include Pasochoa and Cajas.

Cotingas

Andean Cock-of-the-rock *Rupicola peruvianus* 30–32cm

A much sought-after species. Male bright orange (redder in west Andes) with a large bushy crest; black wings have broad silvery patches. Female chocolate-brown, with a much shorter crest. Male's bold plumage used in spectacular display behaviours that includes dancing, jumping, bowing, wing-flapping, chasing and displacing, all while uttering loud, pig-like grunts, chuckles and squeaks. Female nests by waterfalls and rocky slopes.

Where to see Subtropical to foothill forests in Andes, especially forested ravines. Famous leks near Mindo and Baeza.

♀

♂ east

Purple-throated Fruitcrow *Querula purpurata* 25–28cm

Large and conspicuous cotinga that can look all dark. In good light, the deep red-purple throat of male glows. Travels in small groups that maintain constant vocal contact; gives odd, rising *hoouua* calls in series. Bounding flight; shakes tail on landing. Sometimes joins mixed-species flocks of large frugivores.

Where to see Humid forest and borders in north-west and eastern lowlands, including Amazonian lodges with canopy towers, and Chocó sites like Mache, Mashpi and Silanche.

Spangled Cotinga
Cotinga cayana 19–20cm

Male is shining turquoise blue with black spangles and a violet throat patch. Female much drabber, all dark brown with heavy whitish scales on upperparts and dusky scales on paler underparts. Often seen perched lethargically atop trees; sluggish while foraging on fruit. Sometimes found together with Plum-throated Cotinga.

Where to see Canopy of terra firme forest in Amazonian lowlands; regularly seen from canopy towers but also in forest borders.

Long-wattled Umbrellabird *Cephalopterus penduliger* 36–42cm

Unique and spectacular! Glossy black like a crow, but adult male has a very long feathered wattle and large, Elvis Presley-like crest. Female smaller, with much shorter wattle and bushy crest. Courtship display is stunning, males dancing, leaping, leaning forwards, shaking wings and delivering a deep booming call. Otherwise, mostly occurs alone, foraging at large fruiting palms or crossing semi-open areas in undulating flight.

Where to see Confined to humid forests, mostly in north-west lowlands to foothills; two well-known sites exist 2.5 hours from Quito: 23 de Junio and Mashpi–Amagusa.

♂

♀

Bare-necked Fruitcrow *Gymnoderus foetidus* 32–38cm

>Cotingas/Tityras and becards

A large and lanky cotinga uniquely clad in velvet-black, male with grey wings. Also unique are its bluish-grey bill and bare bluish skin on throat and neck giving it a vulture-like appearance. Mostly in small groups, often hopping heavily through canopy, then embarking on long, slow undulating flight.

Where to see Amazonian lowlands in terra firme and inundated forests, river islands and riparian forests. Common in Cuyabeno.

Masked Tityra *Tityra semifasciata* 21–22cm

The attractive tityras were formerly classified as cotingas. Male mostly white, with black mask, black wings and grey tail; rosy-red bill and orbital skin. Female has greyish-brown head and upperparts. Perches sluggishly atop trees, but is regularly more active than canopy cotingas. Moves in small groups that keep vocal contact using odd nasal grunts.

Where to see Humid forest and adjacent clearings in Amazonian and Pacific lowlands, foothills and locally into subtropics.

Tityras and becards

One-coloured Becard *Pachyramphus homochrous* 16–17cm

A large, uniform becard that shares its habitat with other similar species. Getting to know this species will help to identify Slaty and Cinnamon Becards too. Male One-coloured is blackish, crown somewhat darker, wings plain. Female all rufous, darker on upperparts, with a dusky loral area. It gives a sputtering series of thin whistles. Moves boldly in pairs or small groups, often with mixed-species flocks. It often nods its head and stands upright.

Where to see Fairly common in humid to deciduous forests, borders and adjacent clearings in most Pacific lowlands, foothills and subtropics.

♀

♂

Barred Becard *Pachyramphus versicolor* 12–13cm

Males and females are equally eye-catching. Male has a black crown and upperparts, yellow spectacles, cheeks and throat, white breast and belly; the underparts have profuse blackish barring. Female has grey crown, olive upperparts, rufous wings and profusely barred yellowish underparts. Mainly in pairs with mixed-species flocks, hover-gleaning and sallying after escaping prey.

Where to see Cloud forest and borders in the east and north-west Andes; typical in Guacamayos or Bellavista flocks.

White-browed Purpletuft *Iodopleura isabellae* 11–12cm

Small and chunky, cotinga-like bird, with a short bill. Blackish upperparts; white facial markings, rump and underparts. Males have semi-concealed purple tufts on flanks (white in females). Rather sluggish pairs and small groups; perches upright atop bare twigs and snags. Makes aerial sallies after prey, also takes fruit in flight.

Where to see Terra firme forest canopy and borders; often seen from canopy towers at Río Napo lodges and in forest borders at Cuyabeno.

Wing-barred Piprites *Piprites chloris* 12–13cm

A small, smart-looking bird with large eyes, pinkish legs and a bold pale orbital ring. Upperparts greyish olive, nape grey, underparts yellow; two bold yellowish wing-bars and pale yellow edges to flight feathers. Rather lively, often with mixed-species flocks; gleans insects in upper strata. Voice a cadenced and hesitant series of loud notes.

Where to see Terra firme forest and borders; regular at canopy towers at Río Napo lodges and Tamandua, near Puyo.

Scale-crested Pygmy-tyrant *Lophotriccus pileatus* 10–11cm

Tiny tyrant with a characteristic black-and-orange crest often held flat, but sometimes fanned. Eyes whitish. Streaked underparts and bold yellowish fringes to wing feathers. Mostly single birds in shady understorey and mid-strata; makes sudden darts to foliage. Males sing a fast, loud and metallic *treet*.

Where to see Humid to semi-humid forest and borders in Pacific lowlands, foothills and subtropics, and Amazonian foothills and subtropics. Classic Mindo, Sumaco and Ayampe species.

west

Rufous-crowned Tody-flycatcher *Poecilotriccus ruficeps* 9–10cm

A beautiful small flycatcher. Its rufous crown, buffy cheek with black outline, and white throat are distinctive. Has a variable blackish breast-band, rich yellow belly and two yellowish-buff wing-bars. Single birds or pairs are rather unobtrusive, but energetic and vocal. Sallies upwards or forwards to foliage. Gives a brief stuttering trill.

Where to see Montane cloud forest, shrubby edges and bamboo patches in east and west Andes; a default bird in bamboo stands at Cosanga.

east

Common Tody-flycatcher *Todirostrum cinereum* 9–10cm

As its name implies, common and widespread. Its black crown, rich yellow underparts, white throat (only in west) and pale eyes are diagnostic. Mostly found in pairs in semi-open areas; nimble, noisy and rather confiding. Inspects undersides of leaves, tilting to reach them from below. Flicks its cocked tail sideways.

Where to see Humid woodland, shrubby clearings, forest borders and gardens in Pacific lowlands and foothills, and Amazonian foothills and subtropics.

east

Yellow-browed Tody-flycatcher *Todirostrum chrysocrotaphum* 9–10cm

This stunning little flycatcher has a black cap with bold yellow ocular and malar stripes, and white spot in front of eye. Bright olive upperparts, rich yellow underparts, breast with several streaks forming a partial collar. Single birds or pairs in upper strata, making upward sallies to foliage; only rarely with mixed-species flocks.

Where to see Canopy of Amazonian forest, borders and adjacent woodland in the lowlands; regularly seen from canopy towers.

Cinnamon Flycatcher *Pyrrhomyias cinnamomeus* 12–13cm

A smart little flycatcher with rich cinnamon-rufous underparts and olive-brown upperparts; wings have two rufous bars and bold rufous edges to secondaries. Alert pairs, highly sedentary; makes short, sudden loop sallies and often returns repeatedly to the same perch. Rather confiding.

Where to see Cloud forest borders, woodland and adjacent clearings in east and west Andes, up to 3,000m; default flycatcher in Cosanga and Bellavista areas.

Ornate Flycatcher *Myiotriccus ornatus* 12–13cm

Unique, well-named flycatcher, with a distinctive white spot in front of eyes. Slate-grey head, olive chest and back, bright yellow belly and rump, and mostly rufous tail. Sedentary pairs or single birds make sudden loop sallies into air, or to foliage or even ground; perch erect, often at trail edges and forest borders.

Where to see Cloud forest and woodland in east and west Andes, also high coastal ridges; numerous at many frequently visited sites like Mindo.

east

Southern Beardless-tyrannulet *Camptostoma obsoletum* 9–10cm

A lively tyrannulet with a bushy crest, weak facial pattern and short bill. Greyish-olive upperparts, underparts vary from whitish to yellowish. Two buffy wing-bars and short tail. Single birds or pairs, restless and vocal; glean and flit to foliage, also sally with quick movements.

Where to see Humid to dry forest, woodland, arid scrub, clearings and gardens in Pacific lowlands and foothills, temperate Andean valleys, but also forest canopy in the Amazonian lowlands.

Torrent Tyrannulet *Serpophaga cinerea* 11–12cm

Easily identified by habitat and plumage. Has a black cap, wings and tail; ash-grey upperparts, paler underparts; two narrow white wing-bars. Mainly in pairs, lively, regularly flicking tail. Perches on boulders and trunks above rushing water, and makes short aerial sallies to air or ground. Sometimes in slow-flowing channels and dams.

Where to see Andean foothills, subtropics and temperate valleys up to 3,100m; common near towns.

Black Phoebe *Sayornis nigricans* 17–18cm

All sooty-black with white lower belly, white wing-bars, edges to flight feathers and outer tail feathers. Confiding and active; pairs perch on the ground, boulders, posts and higher up on wires and exposed branches. Makes aerial sallies, flicks tail when perched. Often nests under roofs, ledges and other structures.

Where to see Open areas, often near water, in Andean foothills, subtropics and temperate valleys; sometimes in poor-quality rivers and streams.

Vermilion Flycatcher *Pyrocephalus rubinus* 14–15cm

The strikingly coloured male is unique among American flycatchers, coloured as its name implies. More discreet female has faint dusky streaks on breast- and belly-sides, and pinkish wash on lower belly. Bold pairs, sometimes trios, perch on exposed branches and sally into air or to ground after invertebrate prey.

Where to see Common in dry Andean valleys and open and semi-open areas of arid south-west and coast; locally in Amazon. Galápagos has its own species allied to the Vermillion Flycatcher, the Brujo Flycatcher, which is regular on Isabela and Pinzón islands, but rarer elsewhere.

Drab Water-tyrant *Ochthornis littoralis* 13–14cm

As its name implies, it is drab, sandy-brown; darker wings and tail; short, faint, pale eyebrow and narrow eye-ring. Single birds or pairs always low near water, perching on open branches, snags in sandbars and exposed roots in riverbanks. Makes short sallies into air and repeatedly flies ahead of small boats.

Where to see Edges of large and mid-sized, slow-flowing rivers in Amazonian lowlands; easily seen along the Río Napo.

Masked Water-tyrant *Fluvicola nengeta* 14–15cm

A very distinctive snowy-white, longish-legged tyrant, with a narrow black ocular stripe, black wings and black tail with broad white tips. Conspicuous and rather noisy, regularly in pairs or small groups; walks on shore, or floating vegetation in damp areas and their margins. Fans and cocks tail, and droops wings.

Where to see Western lowlands and foothills near water and damp places, including rice fields, shrimp ponds and outskirts of towns.

Rufous-breasted Chat-tyrant *Ochthoeca rufipectoralis* 13–14cm

Chat-tyrants are elegant birds of the forest edge. This species is characterised by its orange-rufous breast patch and long, broad white eyebrow. Belly white, dusky wings with one bold rufous bar. Mostly in pairs, perching upright and well exposed, making sallies into air or to foliage.

Where to see Montane forest and adjacent clearings in east and west Andes up to 3,300m; easy to see along roads that transect the cordillera.

Long-tailed Tyrant
Colonia colonus 20–23cm

This striking tyrant looks absurdly long-tailed. Mostly black, but crown, central back and rump are white. Males have very long tail-streamers, shorter in females. Highly sedentary pairs or small groups perch on exposed branches and stumps. Makes long aerial sallies after flying insects.

Where to see Humid forest borders in north-west lowlands and foothills in east Andes; regular in clearings with trees, like cattle pasture, near forests.

♂ west

Bright-rumped Attila *Attila spadiceus* 18–19cm

This variable bird sports olive, greyish or rufous plumages, all with a yellowish rump, paler belly, vague pale superciliary stripes, and two obvious paler wingbars. A lethargic canopy dweller that often is difficult to locate, unless sallying into air after passing prey; also captures it by hovering at foliage, and eats much fruit. Very vocal, sometimes uttering its leisurely rhythmic song for long periods.

Where to see Regular in humid forest, borders and adjacent clearings in Amazonian and Pacific lowlands and foothills; easily found at lodges along the Río Napo and reserves in north-west Pichincha.

olive morph

rufous morph

Galápagos Flycatcher *Myiarchus magnirostris* 15–16cm

Unmistakeable in Galápagos. Angular head and looks large-eyed. Upperparts plain greyish brown, throat and breast off-white, belly pale yellowish; two pale wing-bars. Tame, even curious; can perch on hats, tripods, cameras and the like. Perches upright and makes sallies into air or to ground.

Where to see Most Galápagos islands, especially in arid lowlands. Easier to see than Brujo Flycatcher, the only other flycatcher species in Galápagos.

Great Kiskadee *Pitangus sulphuratus* 20–21cm

Several flycatcher species share the general pattern of kiskadees and inhabit fairly similar habitats. This species has a heavy bill and its white diadem nearly encircles the whole crown; its dusky tail and wing feathers have rusty edges. Very vocal, energetic and conspicuous; makes long sallies into air, or to ground or foliage. Gives a loud *kiss-ka-dee!*

Where to see
Widespread in open areas in the Amazonian lowlands and foothills, often near water.

Boat-billed Flycatcher *Megarynchus pitangua* 22–23cm

The largest species of the white-diademed group. Resembles Great Kiskadee but has a massive bill, duskier wings and tail, more olive upperparts. Single birds or pairs, loud and conspicuous mainly in upper strata. Makes sudden attacks at foliage, into air or even to ground, seeking largish prey that is smashed against branches before swallowing.

Where to see Humid and deciduous forest borders, canopy and adjacent clearings in the Amazonian and Pacific lowlands and foothills.

Social Flycatcher *Myiozetetes similis* 16–17cm

Social Flycatcher is not especially 'social', but readily identified by its short bill, dull blackish mask, long and broad white eyebrow, olive upperparts and blackish wings with pale feather edges. Mostly in pairs, sometimes alone, in a variety of open areas. Makes long sallies into air, at foliage or to ground.

Where to see Amazonian and Pacific lowlands and foothills, in forest borders, clearings, river edges and towns, often near the similar Rusty-margined Flycatcher.

east

Golden-crowned Flycatcher *Myiodynastes chrysocephalus* 20–21cm

This loud-voiced flycatcher of cloud forests is distinguished by its large bill, greyish head, broad white eyebrow and moustachial, dusky submalar stripe and dusky-olive streaks on a yellow breast. Pairs favour exposed branches. Nods head and perches in a hunched position. Voice loud and whining.

Where to see Forest canopy and borders in foothills and subtropics in east and west Andes, including the Cosanga, Buenaventura, Chical and Bellavista areas.

Streaked Flycatcher *Myiodynastes maculatus* 20–22cm

A bold flycatcher that is densely streaked on both upper- and underparts. Whitish or yellowish eyebrow and moustachial are broad, with black mask; rufous edges to flight feathers. Single birds or pairs favour exposed perches, sometimes loud. Sallies from perch and often chases prey, but also eats much fruit.

Where to see Humid to deciduous forest canopy and borders in the Pacific and Amazonian lowlands and foothills; easier to see and commoner in the Pacific.

east

Snowy-throated Kingbird
Tyrannus niveigularis 18–19cm

This species somewhat recalls the commoner Tropical Kingbird, but has a concolorous back and crown, clear-cut black mask, immaculate white throat and paler grey breast. Single birds or pairs sit atop trees or other exposed perches. Makes long aerial sallies and pursues flying prey. Breeds in drier south-west and moves north after breeding.

Where to see Forest borders and adjacent clearings in Pacific lowlands and foothills; breeds in deciduous forest and scrub of drier south-west, then migrates north to humid forest.

Tropical Kingbird
Tyrannus melancholicus 21–22cm

One of the commonest birds in Ecuador, readily found in many habitats. Grey crown, dusky mask, olive upperparts, whitish throat grading into dull olive breast; tail notched. Single birds, pairs or small groups on wires and other exposed perches, including streetlamps, making long aerial sallies and pursuits.

Where to see Open and semi-open areas and forest canopy in Amazonian and Pacific lowlands, foothills, subtropics and temperate Andean valleys.

Rufous-browed Peppershrike *Cyclarhis gujanensis* 15cm

Peppershrikes are large vireos with a heavy, hooked bill. Broad rufous eyebrow, greyish cheeks, olive upperparts, yellow breast, pink legs and pale horn bill. Pairs skulk in dense foliage and tangles, briefly emerging into open before 'disappearing' again. Utters a rich and melodious song.

Where to see Dry forest and borders in Pacific lowlands, and humid montane forest and borders in south-east Andean foothills. Often seen at Cerro Blanco.

Slaty-capped Shrike-vireo *Vireolanius leucotis* 14cm

This olive and bright yellow vireo has a striking head pattern with a broad lemon-yellow eyebrow, bluish-grey crown and white ornaments below and behind its pale grey eyes. Birds in the Pacific lowlands have pink legs and lack white facial decorations. Mostly in pairs with mixed-species flocks; sluggish, its endless descending whistles are the best indication of its presence.

Where to see Humid forest in the north-west, and Amazonian foothills and adjacent lowlands. Good sites include Mashpi–Amagusa and Tamandua.

east

Brown-capped Vireo *Vireo leucophrys* 12–13cm

Typical vireos are generally drab, but beautiful when seen closely. This species has a brown crown, long white eyebrow, plain wings and pale yellowish underparts. Mostly in pairs with mixed-species flocks; pairs are active, deliberately inspecting foliage and the underside of leaves. Gives a fast, hesitant, warbling song that ends in a rising note.

Where to see Humid forest and borders in Andean foothills and subtropics, common along roads that traverse Andean cloud forests.

west

Chivi Vireo *Vireo chivi* 14–15cm

A common vireo with a complex distribution that includes resident breeding and austral migrant populations. Bluish-grey crown, long white eyebrow narrowly outlined black; olive upperparts, whitish underparts with yellowish on sides and flanks. Eyes dark red. Regularly in pairs, but on migration several might join mixed-species flocks. Gleans foliage, inspects its underside, and attacks prey from perch.

Where to see Humid to dry forests, borders and shrubby clearings in Amazonian and Pacific lowlands, foothills and subtropics.

resident west

Turquoise Jay *Cyanolyca turcosa* 31–33cm

This rather shy jay has a loud, metallic voice and a varied repertoire of whistles, hisses and cries. Mostly turquoise-blue, with silvery crown and throat, and black mask and pectoral collar. Moves in small groups in upper strata, sometimes with other large passerines (caciques, mountain-tanagers). Noisy and wary.

Where to see Montane forest, borders and adjacent clearings up to 3,300m in the Andes; usual sites include Yanacocha and Tapichalaca.

Green Jay *Cyanocorax yncas* 29–32cm

A smart and inquisitive jay that often 'studies' its observers by approaching them. Unique combination of green-and-yellow body, with white head, black face and breast, and blue ornaments on face. Small groups constantly keep vocal contact with a variety of metallic rings, trills, hisses and clicks. Often with oropendolas and caciques.

Where to see Cloud forest, borders and adjacent clearings in east Andes; common in places like San Isidro.

Violaceous Jay *Cyanocorax violaceus* 35–37cm

This large, bold and noisy species is the only jay in the Ecuadorian Amazon. Black hood contrasts with silvery nape, purplish upperparts and pale purplish underparts. Forms small, active and inquisitive flocks that move widely across forested and semi-open habitats, calling loudly. Sometimes joins other large frugivores.

Where to see Amazonian lowlands to foothills, where conspicuous in forest borders, woodland, adjacent clearings and river islands; regularly seen during boat journeys.

White-tailed Jay *Cyanocorax mystacalis* 31–33cm

A striking and smart jay of Pacific dry forests; unique combination of immaculate white and deep blue on body, with black face and breast. Bright yellow eyes and white ornaments on face are also distinctive. Small noisy groups in upper strata, sometimes descend to undergrowth and ground when seeking food or water.

Where to see Dry forest, woodland, arid scrub and even semi-urban areas with natural vegetation in south-west lowlands.

Black-capped Donacobius *Donacobius atricapilla* 21–22cm

White-banded Swallow *Atticora fasciata* 14–15cm

This charming, long-tailed swallow is entirely glossy blue-black with a clean white breast-band. Juvenile is duller. Always found near water in small flocks that perch tightly together on posts, snags, boulders, wires, driftwood, etc; zigzagging flight, circling low above water. As typical of a swallow, groups keep vocal contact as if chattering.

Where to see Slow-flowing rivers, large streams and lagoons in Amazonian lowlands and foothills, both in deep jungle and anthropogenic habitats.

juv.

Grey-breasted Martin
Progne chalybea 18–19cm

Martins resemble large swallows. This species has dark steel-blue upperparts, brownish-grey throat and breast, and dull whitish underparts. Female is duller. Many thousands gather on wires and under eaves at dusk, but by day disperse into smaller feeding flocks that often fly high above ground; long glides and short flaps.

Where to see Common and widespread in eastern and western lowlands and foothills; more abundant near water.

♂

White-winged Swallow
Tachycineta albiventris 13–14cm

Easily differentiated from the other typical swallow of Amazonian rivers and lakes (White-banded) by its glossy greenish-blue upperparts, white rump and large wing patch, and white underparts. Forms small flocks that circle low above water and perch in groups or alone on snags, boulders, driftwood, wires, bridges, etc. Sometimes occurs near White-banded.

Where to see Slow-flowing rivers and lagoons in Amazonian lowlands, both in deep forest and open areas near towns and ports.

Chestnut-collared Swallow
Petrochelidon rufocollaris 12–13cm

A lovely swallow with glossy blue-black crown and upperparts, back with white streaks. Rufous-chestnut front, nuchal collar, 'vest' and rump; white throat and belly. Small foraging flocks circle low above open areas and water. Nesting flocks can be large, especially under bridges and eaves.

Where to see Arid lowlands in south-west, but up to 2,000m in Loja province; default swallow in many towns and cities in Pacific lowlands.

Tropical Gnatcatcher *Polioptila plumbea* 11cm

Attractive little bird with characteristic black, grey and white plumage, and a long tail that is constantly wagged while moving through vegetation. Crown black in males, grey in females. Notable differences in vocalisations and consistent plumage differences suggest that Amazonian and Pacific lowland populations are separate species. Restless, often joins mixed-species flocks, and is seen from forest canopy to low shrubs near coasts.

Where to see Pacific birds are common, especially in deciduous forests and scrub, including on Isla de la Plata; the Amazonian bird is a rarer canopy-dweller.

♀ west

♂ west

Half-collared Gnatwren *Microbates cinereiventris* 10–11cm

A cute little bird with a long slender bill. Its cinnamon face, black collar and very short tail create a distinctive combination. Although not easy to see, patience can be rewarded as this species can be quite tame, often ignoring quiet observers while moving in pairs or small groups low above the ground. It keeps wagging its tail while restlessly changing perches.

Where to see Inside humid forest in Amazonian and Pacific lowlands and foothills.

west

Scaly-breasted Wren *Microcerculus marginatus* 11–12cm

The beautiful song of this species is a common and unforgettable feature of humid forests. It is a fading, descending and lengthy series of high whistles. Its long slender bill, longish legs, tiny tail and whitish underparts with variable dusky scaling are characteristic. Often found in loose pairs on the forest floor, constantly teetering its rear parts.

Where to see Amazonian and Pacific lowlands and foothills, mostly up to 1,100m.

east

House Wren *Troglodytes aedon* 11–12cm

One of our most ubiquitous birds, known to enter houses and believed to clean them of cockroaches and other bugs. This small, dull brown bird has a pale supercilium and dusky barring on wings and tail. Noisy and restless when foraging, almost constantly giving a buzzy call and suddenly singing a fast wheezing melody.

Where to see Cities, towns and open areas in lowlands up to temperate Andean valleys.

Sedge Wren *Cistothorus platensis* 10–11cm

No other wren enters páramo grasslands near the treeline. This small species has a notable streaky pattern above, pale eyebrow and buffy underparts. Somewhat skulking when foraging, but often sings from exposed perches atop small bushes, grass clumps and even logs and posts. Gives a variety of high-pitched notes.

Where to see Andean highlands up to 4,000m in grassy fields and adjacent agricultural areas; common in places like Cajas and Antisana.

Thrush-like Wren *Campylorhynchus turdinus* 20–21cm

This noisy wren is easily recognised by its dull greyish upperparts and whitish underparts with heavy dusky spots, coupled with a long white eyebrow and barred tail. It clambers in pairs or small compact groups high above ground, including the canopy of palms, large limbs and epiphyte clumps, but moves down into vine tangles at forest borders. Utters a rhythmic and cheerful chorus.

Where to see Common throughout Amazonian lowlands and foothills, in forest and adjacent clearings.

Bay Wren *Cantorchilus nigricapillus* 14–15cm

Wrens are gifted songsters, but many are rather dull-plumaged. This species is an exception due to its bright rufous upperparts, contrasting black head and heavy black and white barring below. It prefers wet forest borders, dense thickets and second growth, where it forages in pairs. Its vigorous song has a ringing quality.

Where to see Common in western lowlands and foothills, including well-known sites like Buenaventura, Playa de Oro, Silanche and Palenque.

Superciliated Wren *Cantorchilus superciliaris* 14–15cm

Less skulking than most other wrens, and readily recognised by its bicoloured pattern, long white eyebrow and longish barred tail. Moves in pairs that keep vocal contact; often near ground, but moves higher especially in company of mixed-species flocks. Gives a variety of musical warbles and churrs.

Where to see Arid scrub, deciduous forest and borders, adjacent clearings in south-west lowlands and foothills; easy to see at Cerro Blanco and La Segua.

Grey-breasted Wood-wren *Henicorhina leucophrys* 10–11cm

Small, chunky and almost tailless. Typically found at higher elevations than White-breasted Wood-wren. Its bold facial pattern and grey underparts are characteristic. Busy pairs creep rapidly in forest undergrowth, often near the ground; also in dense thickets and vine tangles. Gives loud and variable musical phrases.

Where to see Cloud forest in east and west Andes, mostly between 1,500–3,000m; common along forested roads that transect the Andes.

White-capped Dipper *Cinclus leucocephalus* 15–16cm

Unique. The only dipper in Ecuador is a very distinctive, chunky and pied passerine tied to rushing streams and rocky rivers, where it searches for aquatic invertebrates. Often in pairs that perch on rocks, boulders and less often on riverside vegetation or logs. Flies low over water with fast flaps.

Where to see Clean-water rivers in east and west Andes, also locally in temperate Andean valleys. Classic sites include Mindo and Cosanga.

Galápagos Mockingbird *Mimus parvulus* 25–26cm

The most widespread of the Galápagos islands' four endemic mockingbird species; does not overlap with any other species on all islands it inhabits. Has a blackish mask, white throat and nuchal collar, dusky-grey upperparts with even duskier streaks, and whitish underparts. In pairs or small family groups; tame, conspicuous and noisy, often perching atop bushes and cacti, but feeding mostly on the ground.

Where to see Most islands except San Cristóbal, Española and Floreana, where other mockingbird species occur.

Long-tailed Mockingbird *Mimus longicaudatus* 28–30cm

The only mockingbird in most of the Pacific lowlands, often in very arid areas. Long white eyebrow, black moustachial, greyish-brown upperparts with darker streaks, bold white on wings, and long tail with broad white tips. Conspicuous, active and noisy, regularly found in groups, giving various whistles, wheezes and gurgles.

Where to see South-west lowlands including coasts and green areas in large cities; also in foothills and subtropical valleys in southern Loja province.

Tropical Mockingbird *Mimus gilvus* 24–26cm

A newcomer to the country, this mockingbird first arrived in northern Ecuador in the late 1990s and is rapidly spreading south, west and east. Does not overlap (yet) with Long-tailed Mockingbird, but is identified by its yellow eyes, plain grey upperparts, pearl-grey underparts and black ocular stripe. Mostly in active and tame pairs or small groups. Prolific songster.

Where to see Temperate valleys in northern Andes; more locally in east and west lowlands and Andean slopes.

Andean Solitaire *Myadestes ralloides* 16–18cm

Solitaires are small forest thrushes. Andean combines grey with rufescent-brown and has silvery white in its flight feathers and outer tail. Rather elusive, more often heard than seen, but patient observers might get a look as it remains on same perch for a long time. A lovely, fluty, ventriloquial song.

Where to see Humid forest and borders in west and east Andes; regular along roads that transect the Andes.

east

Black-billed Thrush *Turdus ignobilis* 22–23cm

Four typical thrush species occur in Amazonian Ecuador; this species is the only one regularly found in non-forest habitats. Dull olive-brown upperparts, greyish-buff underparts with dusky streaks, and a white crescent on throat; bill, eye-ring and legs dark. Gives a soft, pleasant, repetitive song. Often forages on the ground, but sings from high perches.

Where to see Common and widespread in Amazonian lowlands and foothills, including cities and towns.

Plumbeous-backed Thrush *Turdus reevei* 22–23cm

Its bluish-grey upperparts and pinkish-buff underparts are unpretentiously beautiful. The yellow bill and legs, white eyes, grey eye-ring and streaky throat separate from co-occurring Ecuadorian Thrush, which is browner and dark-eyed. Rather tame, often in groups and follows mixed-species flocks. Gives a fast and clear warbling song.

Where to see Deciduous forest, woodland and arid scrub in south-west lowlands, foothills and subtropics. Classic thrush in Cerro Blanco, Jorupe and Machalilla.

Great Thrush *Turdus fuscater* 30–33cm

Our largest thrush is an archetypal urban bird. Its bright orange bill and legs stand out; males have an orange eye-ring too. Walks and runs on the ground, but sings from high perches and forages in various strata too. Its rich, musical song is one of the most beautiful natural sounds in many Andean cities.

Where to see Andean highlands to 4,000m, in forest borders, páramo, agricultural fields, arid valleys, and urban parks and gardens.

♀

House Sparrow *Passer domesticus* 15cm

The only introduced species covered in this book. The black bib and chestnut neck of males are distinctive. Females have dusty-brown upperparts, pale buff underparts, with a faint pale eyebrow. The resident Rufous-collared Sparrow of the Andes has a peaked crest. Gregarious in parks and town squares, never away from urban areas; nests in palms and araucarias.

Where to see Well established in north Andean and lowland cities and towns, including coastal towns.

Lesser Goldfinch *Spinus psaltria* 11–12cm

Recalls Eurasian goldfinches, but male clad in black and bright yellow, with a white patch on wings. Female dull olive above, duller yellow below, but has a similar wing pattern. Moves in pairs or small flocks in various strata; restless and noisy, but less so than the familiar Hooded Siskin.

Where to see Makes rather erratic migratory movements, but is sometimes numerous in agricultural fields, hedgerows and shrubby clearings in the Andes.

Thick-billed Euphonia *Euphonia laniirostris* 11–12cm

A common and widespread euphonia, the Thick-billed is readily identified in male plumage by the rich yellow underparts extending all the way to the chin. Female nondescript like other female euphonias, but has a greyish loral spot. This euphonia is also a prolific imitator and songster; joins mixed-species flocks, and feeds largely on fruit.

Where to see Less forest-based than other tropical euphonias and the only one in deciduous forests; readily seen in gardens and clearings in the Amazonian and Pacific lowlands to foothills.

Golden-rumped Euphonia *Euphonia cyanocephala* 11–12cm

The only truly Andean euphonia is also the most strikingly coloured, both in male and female plumages. Male has golden-yellow rump and belly, deep blue upperparts, sky-blue crown and black mask. The olive female shares the sky-blue crown, but her forehead is orange. Moves in pairs or small groups, sometimes with mixed flocks but more often with conspecifics alone. Feeds mostly on mistletoes, which are 'planted' when the euphonia defecates their sticky seeds.

Where to see Regular in dry Andean valleys around major cities and towns, including Quito, Ibarra, Cuenca and others. Locally also in more humid subtropics.

Orange-bellied Euphonia *Euphonia xanthogaster* 11–12cm

The commonest euphonia in Ecuador. Familiarity with this species will enable observers to identify other euphonias. Male combines orange-yellow with steel blue; its coronal patch is large compared to other euphonias. Female distinguished by her dull orange crown, and greyish nape and breast. This species is highly vocal and a prolific imitator of other birds. Regular in mixed-species flocks; fond of small berries like mistletoes.

Where to see Widespread in forested habitats and adjacent clearings, up to 2,000m. The default euphonia in cloud forests.

♀

♂

Hooded Siskin *Spinus magellanicus* 10–11cm

All siskins are regularly found in flocks; Hooded is the commonest in Andean valleys. Males have a black hood, with rich yellow underparts and a prominent yellow wing patch. Females and young males are greyish olive on upperparts, pale yellowish below, with a similar wing pattern. Gregarious, restless and noisy, their flight is undulating.

Where to see Widespread in the Andes including parks and gardens in cities and towns; also agricultural areas.

Yellow-throated Chlorospingus *Chlorospingus flavigularis* 14–15cm

Chlorospingus are rather dull but energetic birds of humid forest and borders. This species has olive upperparts, greyish underparts and a rich yellow throat. In groups of up to 12, which are active and noisy; regularly joins mixed-species flocks foraging in various strata. Fond of berries and melastomes.

Where to see Foothills and subtropics in the east and west Andes, and in isolated coastal ranges, at sites like Mindo, Bilsa, Narupa, Mera and others.

Yellow-browed Sparrow *Ammodramus aurifrons* 12–13cm

The default sparrow of open areas in Amazonia. Generally dull olive-brown with dusky streaks on upperparts, greyish-white underparts, yellow foreface and eyebrow. Often in small flocks, but also found alone; terrestrial and tame. Gives a high-pitched, buzzy *tic-tzzzz*, often from an exposed perch.

Where to see Widespread in grassy and shrubby fields, including agricultural land, cattle pastures and even barren terrain in Amazonian lowlands and foothills.

Orange-billed Sparrow *Arremon aurantiirostris* 15–16cm

The only sparrow with a bright orange bill is also characterised by a striped head pattern, black pectoral band and bright yellow wing bend. Mainly found hopping in pairs on the ground or in dense undergrowth; sometimes visits artificial feeders or briefly emerges from cover, and only seldom occurs with mixed-species understorey flocks.

Where to see Humid forest, borders, adjacent shrubby thickets and agroforestry fields in the Pacific and Amazonian lowlands and foothills.

east

Rufous-collared Sparrow
Zonotrichia capensis 13–14cm

One of the most common and familiar birds in the Andes. This charismatic sparrow has a peaked crest, striped head and rufous nuchal collar. Upperparts dull brown with dusky streaks, and wings have two white bands. Juvenile heavily streaked. Single birds, pairs or small groups hop boldly on the ground or in low vegetation; tame and smart. Sings from exposed perches.

Where to see Andean agricultural fields, towns, gardens, parks and similar habitats.

Yellow-breasted Brushfinch
Atlapetes latinuchus 17–18cm

Brushfinches recall large sparrows but are more forest-dependent. The rich yellow underparts of this species are fairly distinctive, combined with the rufous crown and dark grey upperparts. Pairs or small sprightly groups occur in dense thickets and shrubby borders, mainly away from mixed-species flocks, venturing out of cover briefly.

Where to see Andean highlands, including forests and shrubs above temperate Andean valleys, and creeks in urban and semi-urban areas.

west

White-winged Brushfinch *Atlapetes leucopterus* 15–16cm

This species is often found in close proximity to Yellow-breasted Brushfinch, but readily distinguished by its whitish underparts, white patch in wings, narrow black moustachial and greyer upperparts. South-western birds have variable amounts of white in the face. Also in pairs or small groups, often with mixed-species flocks. Behaviour similar to Yellow-breasted.

Where to see Arid scrub, woodland and shrubby forest borders in the west Andes and slopes above temperate Andean valleys; common in western Loja province.

north

Russet-backed Oropendola *Psarocolius angustifrons* 35–48cm

Oropendolas have spectacular, unmatched voices. The most widespread oropendola is rufescent brown, with a black bill in most of Amazonia, but a yellow bill in the south-east and west Andean slopes. This noisy species often forms large aggregations, especially at dusk. Builds huge hanging nests; display song from exposed perches is an accelerating and liquid gurgling while bowing down.

Where to see Forest canopy, borders and adjacent clearings in Amazonian lowlands and foothills, also locally in western Andean foothills.

east

Scarlet-rumped Cacique *Cacicus uropygialis* 21–29cm

This cacique can appear all black until revealing its bright scarlet rump. Eyes blue, bill yellowish. Travels in small flocks, regularly with other large passerines, including jays, fruitcrows and oropendolas. Builds smaller hanging baskets than oropendolas, and sometimes forms small colonies.

Where to see Humid forest and borders in western lowlands and foothills (Mashpi, Playa de Oro and elsewhere); also in foothills and subtropics on east Andean slope.

east

Yellow-rumped Cacique *Cacicus cela* 24–29cm

The prolific vocal repertoire of this talented mimic can include artificial sounds like car alarms and whistles. Has yellow on rump and tail, blue eyes and yellowish (Amazon) or bluish (Pacific) bill. In noisy flocks, often associated with oropendolas even when nesting; nests also associated with those of wasps.

Where to see Forest borders and adjacent clearings in Amazonian lowlands; semi-humid to deciduous forest in western lowlands to foothills, including urban parks.

east

Oropendolas, orioles and blackbirds

Peruvian Meadowlark *Leistes bellicosus* 20–21cm

A classic open-area bird that is apparently spreading due to habitat conversion. Male has an elegant bright red chest and long whitish eyebrow; female more subdued overall, with a pink wash on chest. Males perform an aerial display during which the fast buzzing song is delivered; they also sing from exposed perches like fence posts. Otherwise, it is primarily terrestrial.

Where to see Fairly common in arid Pacific lowlands and southern Andean valleys up to 2,500m; a ubiquitous species of agricultural land near Guayaquil and Loja cities, and along coasts.

♀

♂

Yellow-tailed Oriole *Icterus mesomelas* 20–22cm

The orioles all have rich, melodious voices. Yellow-tailed is clad in bright yellow and velvet black, has narrow white edges to the flight feathers and yellow in tail. The similar White-edged Oriole has white in outer tail feathers and prefers drier habitats. Orioles are conspicuous when singing atop perches; otherwise, they forage rather unobtrusively.

Where to see Humid forest borders, woodland, second growth, gardens and hedgerows in Pacific lowlands to foothills, including outskirts of cities and towns.

Orange-backed Troupial
Icterus croconotus 23–24cm

A very striking bird of semi-open areas. Its orange eyes and bright blue orbital skin stand out in a black mask and bib; otherwise mainly bright orange, but wings black with a bold white patch. Seen mainly in pairs foraging in dense thickets, briefly emerging onto bare branches. Sings from exposed perches.

Where to see Second growth, lake margins and riparian woodland in Amazonian lowlands; regularly seen in traditional crops near towns and settlements.

Shiny Cowbird *Molothrus bonariensis* 20–22cm

Cowbirds are brood parasites; females lay their eggs in other species' nests and the nestlings are then tended by the host parents to the expense of their own offspring. The glossy purplish-black males are noisy and bold around anthropogenic habitats, often seen harassing other passerines in pursuit of a nesting pair to parasitise, with the dull brown to greyish brown female waiting around to lay her eggs. Seeing the dull brown juveniles being fed by smaller 'parents' (sometimes half the size) is incongruous.

Where to see Common in Pacific and Amazonian lowlands, now spreading to temperate Andean valleys.

♀ south-west

♂

Great-tailed Grackle *Quiscalus mexicanus* 33–46cm

This human-commensal bird, widespread elsewhere in the Americas, is primarily confined to coasts and adjacent areas in Ecuador. Male glossy purple, with bright yellow eyes and a very long 'creased' tail. Female and juvenile male smaller, mostly dull brown, darker on upperparts, with shorter tail. Often seen scavenging on ground on tidal flats, at edges of mangrove and estuaries, but also on crowded popular beaches. Very vocal.

Where to see Unmistakeable and easy to see along the entire coast, and 'omnipresent' in Guayaquil's urban parks and gardens.

Scrub Blackbird
Dives warczewiczi 22–24cm

This familiar human-commensal bird is all black, with a longish tail and pointed bill. Might recall the equally common Shiny Cowbird, but has longer legs, longer bill and duller plumage. Conspicuous, noisy groups, but also single birds and pairs; forages on the ground, but sings from exposed perches. Song very loud and penetrating.

Where to see Open areas in western lowlands to foothills; also more locally in subtropics and temperate Andean valleys.

Oriole Blackbird
Gymnomystax mexicanus 27–31cm

This bold blackbird combines bright golden yellow with black, and has a large, pointed bill. Regularly seen on sandbanks along large Amazonian rivers, where found in pairs or small loose flocks. Forages mainly on the ground, but perches atop bushes, logs and posts to sing.

Where to see Mostly on river islands, open areas and sandbars along large rivers like the Napo, Aguarico and Pastaza in the Amazonian lowlands.

Tropical Parula *Setophaga pitiayumi* 10–11cm

This small and attractive warbler is readily recognised by its greyish-blue crown and wings, olive back, two white wing-bars, orange throat and breast on otherwise yellow underparts and white lower belly. A restless and sprightly bird that often joins mixed-species flocks while foraging in outermost twigs and foliage in upper strata.

Where to see Humid forest and borders in western lowlands to subtropics and foothills and subtropics of east Andes.

Yellow Warbler *Setophaga petechia* 12–13cm

Two forms occur in Ecuador: a breeding resident on coasts, including the Galápagos, and a boreal migrant in continental forests and woodland. Bright yellow with an olive back; male of resident form has reddish crown and reddish breast streaks. The migrant form often has a few faint streaks. Forages actively in foliage and outermost twigs.

Where to see Mangrove and borders (resident), humid forest borders and woodland in lowlands and temperate Andean valleys (migrant).

♂ resident

Blackburnian Warbler *Setophaga fusca* 12–13cm

The commonest boreal-migrant warbler varies in plumage from a stunning bright orange face and breast, with contrasting head and wing patterns (typical breeding male), to dull yellowish face and breast, and a dull wash overall (non-breeding). First birds arrive from North America in October and the last depart by April. Often seen alone, but sometimes a few join mixed-species forest flocks. Energetic while foraging, seemingly always on the move.

Where to see Mostly seen in Andean cloud forests, but also temperate woodland and Andean valleys, including city parks in Quito.

♂ br.

non-br.

Buff-rumped Warbler *Myiothlypis fulvicauda* 13–14cm

A highly distinctive and mainly terrestrial warbler. Upperparts greyish-olive, underparts buff, as is its bold eyebrow, rump and basal half of tail. Mostly in pairs or small family groups, always near water; hops on or near the ground, fanning its tail and moving it sideways.

Where to see Sluggish and fast-flowing streams and damp areas in humid forest and borders in Amazonian and Pacific lowlands and foothills.

west

Russet-crowned Warbler *Myiothlypis coronata* 13–14cm

Its antiphonal duets are among the most charming bird songs of Andean cloud forests. Grey head with black lateral stripes and orange crown are very distinctive. Underparts vary from yellow in north-western populations to grey or greyish-white in east and south-west. Travels in pairs or small noisy groups, often with mixed-species flocks near the ground.

Where to see Forest undergrowth and borders in the east and west Andes, up to 3,300m elevation.

north-west

Canada Warbler *Cardellina canadensis* 13–14cm

One of the commonest boreal-migrant warblers in forested habitats, Canada is identified by its grey upperparts, bold yellow spectacles and yellow underparts with a broken band of blackish streaks on breast. Its longish bill and legs are pink. Single birds or pairs often join mixed-species flocks, nimbly foraging in dense foliage; cocks and fans tail almost permanently.

Where to see Humid forests and borders in Andean foothills and sub-tropics, locally in adjacent lowlands; common in Amazonian foothills.

♂ br.

Slate-throated Redstart *Myioborus miniatus* 13–14cm

This lively arboreal warbler has a slate-grey hood and upperparts, yellow belly and white undertail. Single birds, pairs or small groups join mixed-species flocks; hyperactive as gleans foliage and small twigs. Flicks and fans tail, droops wings and makes sudden sallies.

Where to see Humid forest and borders in foothills and subtropics of east and west Andes; locally in coastal ranges and temperate Andean valleys, including semi-urban parks and gardens near Quito.

Spectacled Redstart *Myioborus melanocephalus* 13–14cm

This lovely and lively forest bird has bold yellow spectacles that in some individuals in north extend onto yellow face and frontal area. Its chestnut crown, and white vent and most tail feathers, are also characteristic. Appears restless when foraging, moving in and out of vegetation in short sallies into air or to foliage, also gleaning foliage and constantly fanning its tail to flash the white.

Where to see Fairly common in Andean forests up to 4,000m; classic member of mixed-species flocks at sites like Papallacta, Cajanuma, San Isidro or Pasochoa.

north

Golden Grosbeak *Pheucticus chrysogaster* 21cm

A rich-yellow, chunky bird of gardens, hedgerows and agricultural fields, also found in deciduous woodland and arid scrub near coasts. Its wings are black with white 'ornaments'. Female has coarse dusky streaking on upperparts. Single birds, pairs or trios are often found high in fruiting trees, but is also fond of maize, sunflower and other crops.

Where to see In south-west lowlands and temperate Andean valleys up to 3,500m; now spreading to deforested Andean slopes.

♂

Summer Tanager *Piranga rubra* 17–18cm

This common boreal migrant shows notable variation in plumage colour. Adult male all red but immature males are greenish yellow, variably blotched orangey or red. Female and younger male greenish yellow, yellower on underparts. Can be numerous locally, otherwise found in pairs often accompanying mixed-species flocks.

Where to see Humid forest borders to gardens and city parks from lowlands to temperate Andean valleys up to 2,800m elevation.

♀

♂

Magpie Tanager *Cissopis leverianus* 25–27cm

A highly distinctive tanager reminiscent of a Eurasian Magpie. Long-tailed, black and white, with bright yellow eyes. Moves in pairs or small flocks, often away from mixed-species flocks; resembles a jay when foraging. Gives several loud metallic notes when crossing open areas. Perches on exposed branches.

Where to see Widespread in Amazonian lowlands and foothills, in terra firme forest and clearings with trees; regular at Río Napo lodges, and around Tena and Puyo.

Fulvous Shrike-tanager *Lanio fulvus* 17–19cm

This stocky tanager has a heavy bill with a hooked tip. Male has black hood, wings and tail, and an ochraceous body. Female has dull olive-brown upperparts, paler underparts, wings more rufescent, crown darker. Mainly in pairs that lead mixed-species flocks, acting as sentinels. Makes rapid sallies after prey.

Where to see Canopy of terra firme forest and borders in Amazonian lowlands and foothills; typically seen from canopy towers.

Red-capped Cardinal *Paroaria gularis* 16–17cm

This beautiful cardinal is readily identified by the bright red head contrasting sharply with a black-and-white body. Juveniles have a similar pattern, but are duller; black of upperparts replaced by dull brown and red of head by buffy-brown. In pairs or small groups, low above water at edges of sluggish rivers, streams, lagoons and flooded forest. Tame and noisy when foraging and perches on exposed branches.

Where to see Fairly common in aquatic ecosystems in the Amazonian lowlands; Cuyabeno, Limoncocha, Añangu, Taracoa and other lagoons are typical sites for it.

juv.

White-shouldered Tanager *Loriotus luctuosus* 13–14cm

A notably dimorphic tanager of humid to deciduous forest and borders. Black male has striking white shoulder patches. Female more confusing, but silvery base to lower mandible distinctive. Combination of grey head, white throat and rich yellow belly, as well as accompanying males, are good identification clues. Regular member of mixed-species flocks, often in hyperactive pairs or small groups that forage mainly in outer foliage, also sallying after prey.

Where to see Fairly common in forested habitats in lowlands and foothills of east and west. One of the few typical tanagers found in dry forests.

Silver-beaked Tanager *Ramphocelus carbo* 17–18cm

One of the commonest Amazonian birds. Male has a silvery base to lower mandible and is entirely wine-red, but can look all dark in poor light. Female reddish brown, but redder on underparts. In pairs or small active groups, mainly in shrubbery, second growth and forest borders. Noisy and energetic.

Where to see Open and semi-open areas in Amazonian lowlands and foothills, including urban parks, but also forest borders.

Masked Crimson Tanager *Ramphocelus nigrogularis* 18–19cm

A stunning tanager of river edges and forest borders, with velvet-black mask contrasting with blood-red body; wings, tail and lower belly also black. Male has silvery base to lower mandible; female only slightly drabber. Forms active groups, sometimes foraging in thickets low above water. May consort with Silver-beaked Tanager.

Where to see Amazonian lowlands largely below 600m, mainly near creeks, streams, rivers and lagoons, but sometimes in second growth.

Flame-rumped Tanager *Ramphocelus flammigerus* 18–19cm

One of the commonest tanagers of forest borders and adjacent clearings in western Ecuador. Male is stunning. Its velvet-black plumage is relieved by a striking lemon-yellow rump and silvery bill. Female shares lemon-yellow rump, but has dusky-brown upperparts and yellow underparts. Conspicuous and noisy even in heat of midday. In pairs or small groups, sometimes with mixed-species flocks, but often outnumbers other species in flocks. Males display their bright rumps from exposed perches.

Where to see Common and widespread in western humid forests up to 2,000m elevation.

♀

♂

Hooded Mountain-tanager *Buthraupis montana* 22–23cm

Heavy and conspicuous; black hood, blue upperparts and yellow underparts; bright red eyes stand out even in dim light. Moves in noisy groups that hop vigorously along branches in a jay-like fashion. Sometimes joins other large birds in mixed flocks, including jays and caciques.

Where to see Cloud forest in east and north-west Andes, up to 3,200m. Regularly seen at Yanacocha, Guacamayos and along roads transecting the Andes.

Grass-green Tanager *Chlorornis riefferii* 20–21cm

Unique and stunning, but sometimes overlooked in foggy weather. Bright green and red, with red bill and legs. Forms small groups and is regular in mixed-species flocks; sometimes conspicuous and vigorous while peering around for invertebrate prey and berries. Gives a characteristic series of nasal *queek* notes.

Where to see Cloud forests and borders in east and north-west Andes, like Hooded Mountain Tanager. Typically seen at Tandayapa, San Isidro, Tapichalaca and many other well-known sites.

Scarlet-bellied Mountain-tanager *Anisognathus igniventris* 18–19cm

This handsome tanager combines velvet-black and scarlet, with blue shoulders and rump, and a scarlet cheek adornment. Noisy and energetic, regularly found in pairs or small flocks either with mixed flocks or alone. Gives a loud, tinkling, semi-musical song. Fond of berries and invertebrates, but also takes occasional flowers and buds.

Where to see Common in montane forest, borders, shrubby clearings and hedgerows in the Andes, including woodland above agricultural valleys.

Blue-winged Mountain-tanager *Anisognathus somptuosus* 18–19cm

The bright lemon-yellow crown and underparts, and blue shoulder patch form a distinctive combination. Back is green in east Andes, but black in west Andes. Moves in noisy groups, regularly with mixed-species flocks for which it apparently acts as a leader. Frequent at artificial feeders.

Where to see Cloud forest and borders in east and west Andes; frequent at popular destinations like Mindo, Bellavista and Amagusa.

north-west

Blue-and-yellow Tanager *Rauenia bonariensis* 16–18cm

One of the authors' favourite garden birds. Male's sky-blue hood, golden-yellow rump and underparts, and green back are characteristic. Female and juvenile male duller, but have a bluish wash on crown, wings and tail. Underparts yellowish-buff. Fond of cultivated fruiting trees, also found in pairs or small groups in dry woodland and scrub; sometimes noisy and bold.

Where to see Fairly common in dry Andean valleys up to 3,000m, from the Colombian border south to Azuay province. Regular in green urban areas of Quito, Ibarra, Cuenca and other cities.

Tanagers

Blue-grey Tanager *Thraupis episcopus* 15–17cm

A very common and widespread tanager, mostly pale bluish, upperparts darker, wings and tail bluer. Birds in Amazonia have a bold white patch on wings. A familiar species of human-modified habitats, sometimes quite tame at feeders, in gardens or cultivated fields; often seen in pairs or small groups.

Where to see Many types of more or less wooded habitats in east and west lowlands, subtropics and temperate Andean valleys, including parks and gardens in cities and towns.

east

Palm Tanager *Thraupis palmarum* 16–17cm

The 'dark' version of Blue-grey Tanager is clad in beautiful olive to greyish-olive, with a darker back and brighter wing-coverts. Behaviour similar to that of Blue-grey Tanager with which it often consorts. Flocks are compact and noisy; regularly joins mixed-species flocks and visits artificial feeders. Fond of palm fronds.

Where to see Common and widespread in Amazonian and Pacific lowlands to subtropics, in forest canopy, borders, agricultural land, gardens and hedgerows.

east

Spotted Tanager *Ixothraupis punctata* 13–14cm

This stunning, well-named tanager is mostly bright green, whiter below, all attractively spotted; head has a bright bluish tinge, edges of wing feathers are bright green. Pairs or small flocks regularly join mixed-species flocks, actively gleaning mossy branches, outer foliage and twigs; fond of berries.

Where to see Cloud forest and borders in foothills and lower subtropics in east Andes; often seen at Sumaco, Puyo, Narupa, Macas, Zamora, Lumbaqui and Tena.

Golden-naped Tanager *Chalcothraupis ruficervix* 13–14cm

All 'typical' tanagers are stunning. This species is predominantly turquoise-blue, with a blacker head, heavy black spotting on upperparts, buffy-white lower belly and vent, and a golden patch on nape. Regularly joins mixed-species flocks, peering along large limbs, inspecting underside of branches, gleaning outer foliage and sallying after prey.

Where to see Cloud forest and borders in Andes, including artificial feeders at Amagusa, Paz de las Aves and other localities.

west

Blue-necked Tanager *Stilpnia cyanicollis* 13–14cm

This common tanager stands out even at distance due to its bright blue hood. Birds in western Ecuador have greener wings and a turquoise rump; eastern ones have golden-green patches and a yellow-green rump. In the Amazonian foothills, Masked Tanager is similar, but its wings are greenish-blue. Blue-necked forms small flocks but also moves in pairs; feeds largely on small berries and buds, and often visits fruit feeders.

Where to see Common in secondary forest, clearings and forest borders in western lowlands to subtropics, e.g. at Mindo, and in eastern foothills to subtropics, e.g. at Puyo.

west

east

Beryl-spangled Tanager
Tangara nigroviridis 13–14cm

A very handsome tanager, with metallic greenish-blue spangles on a black background, and a brighter, more opalescent head, with a black mask, black back and bright blue 'ornaments' on wings. In pairs or small groups with mixed-species flocks, often gleaning outer foliage and bare twigs, and inspecting underside of foliage. May visit artificial feeders at some sites.

Where to see Cloud forest and borders in Andes, including places like Tandayapa and Guacamayos.

Turquoise Tanager *Tangara mexicana* 13–14cm

Its misleading name does not impinge on its beauty: there is no a hint of turquoise in its metallic blue-and-black plumage, with a yellow central belly and shoulders. Mainly in small flocks in upper strata, only briefly joining mixed-species flocks. Inspects small bare branches, epiphyte clumps, dead twigs and mosses, sometimes leaning down; also visits flowers.

Where to see Amazonian forests in lowlands and foothills; best enjoyed from canopy towers.

Paradise Tanager *Tangara chilensis* 14–15cm

Can this tanager be any more beautiful? Its multicoloured appearance includes a lime-green head, purple throat, turquoise-blue underparts and scarlet rump. Active and vivacious groups in upper strata, join mixed-species flocks but also forage away from them. Fond of berries, also inspects bromeliads, outer foliage and moss clumps.

Where to see Forest canopy, borders and adjacent clearings, including towns, in Amazonian lowlands and foothills. Regularly seen from canopy towers.

Bay-headed Tanager
Tangara gyrola 13–14cm

A very distinctive tanager with brick-red head, green upperparts, and blue underparts and rump. Birds in the Amazon have a golden nape and yellow shoulder patch. Small family groups often join mixed-species flocks, but also occur away from them. Works bare branches and twigs, often hanging upside-down to inspect underside. Eats much fruit, especially berries.

Where to see Humid forest, borders, adjacent clearings and gardens in Pacific and Amazonian lowlands, foothills and subtropics.

east

Golden Tanager *Tangara arthus* 13–14cm

This aptly named tanager is rich golden orange with a bold black ear patch; back and wings black with bright golden-orange streaks and feather edges, respectively. Often outnumbers other tanagers in mixed-species flocks. Forages on bare and mossy branches and twigs, regularly inspecting their undersides; also enters outermost foliage.

Where to see Forest and borders in Andean foothills and subtropics, regularly seen at artificial feeders at Mindo, Amagusa, Buenaventura, Copalinga and other sites.

west

Yellow-bellied Dacnis *Dacnis flaviventer* 12–13cm

Male very distinctive, clad in black and yellow, with red eyes and green crown. Female shares red eyes, but has dull olive upperparts, olive-yellow underparts and faint dusky streaks on breast. Nimble pairs regularly join mixed-species flocks, gleaning foliage for invertebrates, but also pecking berries and sipping nectar.

Where to see Fairly common in Amazonian lowlands and foothills, mainly in tree crowns, borders and adjacent clearings; usual seen from canopy towers.

♂

Purple Honeycreeper *Cyanerpes caeruleus* 10–11cm

Honeycreepers, dacnises and flowerpiercers

There are three honeycreeper species in which males are deep purple. This, the commonest and most widespread, has yellow legs and a small black bib. Female has green upperparts, a broad bluish moustachial, and underparts with green streaks. A regular member of mixed-species flocks; often in pairs or small groups, nimble and restlessly gleaning for invertebrates in foliage, also taking small berries and a lot of nectar.

Where to see Humid forest canopy, borders and adjacent clearings in Amazonian and Pacific lowlands and foothills; often seen in canopy flocks.

♀

♂

Green Honeycreeper *Chlorophanes spiza* 13–14cm

The bold male is all bright turquoise except for its black hood. The female is all green, but still distinctive given its longish curved bill. This honeycreeper is also a classic mixed-species flock member, but is also found away from them in pairs and small groups. Also regular at nectar and fruit feeding stations. Feeds largely on nectar and small berries, but invertebrates form much of its diet.

Where to see Fairly common in lowlands to foothills in east and west Ecuador, in humid forest, borders and adjacent clearings.

Glossy Flowerpiercer *Diglossa lafresnayii* 14–15cm

Bill of flowerpiercers is uniquely upturned and hooked, for piercing bases of corollas to reach nectar. This species is all glossy black with a pale bluish patch on shoulders. The similar Black Flowerpiercer lacks a shoulder patch. Single birds or pairs actively cling from flowers, but also glean invertebrates from foliage.

Where to see Montane forest borders, treeline and shrubbery in east and west Andes, up to 3,500m. Common at sites like Yanacocha.

Masked Flowerpiercer *Diglossa cyanea* 14–15cm

The commonest of the blue flower-piercers is also the species that occurs at higher altitudes. It has striking red eyes; all rich blue with a contrasting black mask. Forms small noisy groups that often accompany mixed-species flocks. Actively pecks berries, pierces flowers and probes foliage.

Where to see Montane forest, borders and shrubby clearings in east and west Andes along roads like Nono–Tandayapa or Gualaceo–Limón; locally in urban parks and gardens.

Swallow Tanager *Tersina viridis* 15–16cm

A large and colourful tanager that makes aerial sallies and chases prey; can resemble a large swallow. Bright turquoise-blue male is unmistakeable. Female also very distinctive given its bright green plumage and yellow central belly. They perch atop exposed branches from where they sally into air. Several individuals often congregate at berry-laden trees and then 'disappear', which suggests that they engage in seasonal or erratic migratory movements.

Where to see Uncommon in east and west lowlands to foothills, where seen in humid forest and borders; regularly seen from Amazonian canopy towers.

♂

♀

Plumbeous Sierra-finch *Geospizopsis unicolor* 15–16cm

The default finch of highest páramos, sometimes even in barren rocky areas well above the treeline. The chunky male is all slate-grey. Female densely streaked all over, darker on upperparts, with plain grey rump. The smaller Ash-breasted Sierra-finch occurs at lower elevations and in drier habitats, and is paler. Sierra-finches are terrestrial, often feeding on roadsides, recently ploughed fields, and skulking rapidly in grass clumps. Regularly perches atop boulders and stumps.

Where to see Grassy and shrubby páramo above 3,000m elevation and as high as 4,800m.

Collared Warbling-finch *Poospiza hispaniolensis* 12–13cm

Recalls a sparrow. Its bold white eyebrow, black mask and incomplete black collar are diagnostic, as are its grey sides, white belly and throat. Upperparts grey to brownish. Bill grey to yellowish. Moves by hopping in rather tame pairs or small groups on the ground or in low bushes.

Where to see Arid scrub and woodland in the south-west lowlands and arid valleys in Loja province; common bird on Isla de la Plata.

Saffron Finch *Sicalis flaveola* 13–14cm

Saffron Finch is spreading in the Ecuadorian Andes, Amazonian lowlands and humid north-west. Adult has bright yellow underparts, with rich orange face and front, and olive-yellow upperparts. Juvenile much duller, mostly greyish buff, but yellower on underparts and rump; back has faint dusky streaks. Forages in pairs to small flocks on the ground, often around habitation, cultivated fields, urban gardens and the like. Nests in holes.

Where to see Common, sometimes even numerous, especially in western lowlands and dry Andean valleys, including large cities.

juv.

Blue-black Grassquit *Volatinia jacarina* 10–11cm

Small, finch-like bird readily identified by its sharp bill. Male is glossy blue-black, but can look black in poor light. Female has drab brown upperparts, whitish underparts, coarse streaks on chest to belly-sides. Juvenile male dusky, variably blotched blue black. Males perform a strange display from posts and other exposed perches, leaping up and singing a buzzy *dzzziir*. Otherwise, found in small flocks, often with other seed-eating species.

Where to see Common in grassy fields and shrubby clearings in Amazonian and western lowlands and foothills, also locally in dry temperate Andean valleys.

♂

♀

Parrot-billed Seedeater *Sporophila peruviana* 11–12cm

This highly distinctive seedeater of arid land has a swollen yellowish bill. Male has a bold facial pattern, white bars in wing-coverts and a white speculum in flight feathers; female duller but has similar wing pattern. Can congregate in largish flocks, often with other seed-eating species like Chestnut-throated Seedeater and Blue-black Grassquit.

Where to see Shrubby pastures and arid scrub in south-west lowlands, including the xeric Santa Elena Peninsula and scrub near coasts.

Chestnut-bellied Seedeater *Sporophila castaneiventris* 10–11cm

This common Amazonian seedeater is the easiest to identify in the region, but only in male plumage: slate grey above, chestnut below. Females, on the contrary, are nearly inseparable from other seedeaters. It is advisable to search for males, even if female-plumage birds outnumber males. As is typical of seedeaters, Chestnut-bellied often consorts with other species in grassy and shrubby fields; sometimes flocks can be fairly large and bold.

Where to see Open areas in Amazonian lowlands to foothills, including river islands and edges of large rivers and lagoons.

Chestnut-bellied Seed-finch *Sporophila angolensis* 12–13cm

Seed-finches resemble seedeaters, but are larger and heavier-billed. Male Chestnut-bellied is the only bicoloured seed-finch. Female is chocolate brown and large billed, but bill is smaller than other seed-finches. This species forms small loose flocks or forages alone, but also consorts with other seed-eating birds at seeding herbs and shrubs. Singing individuals perch atop bushes and other exposed perches, and deliver a sweet jumble of twitters and trills.

Where to see Fairly easy to see in Amazonian lowlands to foothills, more often in shrubbery and woodland than in open pasture.

Yellow-bellied Seedeater *Sporophila nigricollis* 10–11cm

Male Yellow-bellied has olive-brown upperparts, a blackish hood and pale yellow on underparts. Bill often looks bluish grey. Female seedeaters can be almost impossible to identify. Found in pairs or small flocks in shrubby fields, but also congregates with other seedeaters in largish groups.

Where to see Widespread in Pacific lowlands, foothills and subtropics, also in temperate Andean valleys near Quito and other cities, and in foothills and subtropics in the east Andes.

Variable Seedeater *Sporophila corvina* 11–12cm

This common seedeater is easily recognised by the male's black-and-white plumage (black collar and white rump are distinctive features), but female is easily confused. The latter is yellowish olive or yellowish buff, but is always yellower than other seedeater species. Forms fairly large flocks in grassy fields and parks, but also frequents shrubby edges.

Where to see Non-forest habitats throughout Pacific lowlands and foothills.

Bananaquit *Coereba flaveola* 10–11cm

A very distinctive small bird with a unique slender, short, curved bill. Readily identified by its bold head pattern, rich yellow rump and mostly yellow underparts, white lower belly and grey throat. Very active and restless, often in fruiting trees pecking at berries or inspecting flowers for nectar; also joins mixed-species flocks.

Where to see Common in humid forest borders, woodland, adjacent clearings and gardens in Pacific lowlands and foothills; also in Amazonian foothills.

Yellow-faced Grassquit *Tiaris olivaceus* 9–10cm

This seedeater-like species has a conical pointed bill, unlike the typical seedeater bill. The striking yellow facial pattern is unique and stands out against the deep black head and breast. The duller females have a similar but subdued pattern. Single birds, pairs or small flocks often perch low in grasses and bushes, hanging from grassy stems, bending them over. Perches high to sing.

Where to see Mostly in north-west foothills and subtropics, in grassy and shrubby fields.

Woodpecker Finch *Camarhynchus pallidus* 14–15cm

Although not remarkably coloured, Galápagos finches are evolutionary benchmarks. Woodpecker Finch is stocky and drab, with a longish conical bill. Its upperparts are greyish, nearly plain, with a whitish crescent above the eyes. Underparts buffy white with duskier streaks on breast. Active pairs in trees; climbs, clings and pecks, and often uses tools to take prey from holes.

Where to see Mainly in highlands of largest islands including Santa Cruz, Isabela and San Cristóbal.

♂

Small Tree-finch *Camarhynchus parvulus* 10–11cm

This small finch has a short stubby bill and short tail. Males have a black hood, pale olive upperparts and dull breast streaking. Females are dull greyish olive above, with a very short whitish eyebrow and faint buff wingbars. Active pairs occur in trees, gleaning foliage and the underside of twigs and leaves; also in flower and fruit clumps.

Where to see Mainly in humid highlands of largest Galápagos islands, but also in woodland near coasts.

♀

Small Ground-finch *Geospiza fuliginosa* 10–11cm

The smallest terrestrial finch is one of the most abundant birds in Galápagos. It has a small triangular bill and short tail. Males are black, with white undertail-coverts. Females are very streaky on the underparts and have a pale bill; the greyish upperparts have dense spotting. Opportunistic, might take ectoparasites from tortoises; takes seeds from ground or picks food scraps from people.

Where to see All of the Galápagos islands except the remote Darwin, Wolf and Genovesa.

Large Ground-finch *Geospiza magnirostris* 15–16cm

The bill can be impressive and is heavily swollen. A thickset, strong-necked species; males mostly black, females heavily streaked and mottled. Mainly in small flocks, sometimes with other terrestrial finches. Cracks hard and large seeds with its strong bill, but also takes buds, invertebrates, cacti flowers and fruit.

Where to see Mostly in arid lowlands of several Galápagos islands, including Fernandina, Isabela, Santiago and Santa Cruz.

Medium Ground-finch *Geospiza fortis* 12–13cm

The default confusing finch. As its name implies, it lies midway between Small and Large Ground-finches. Its triangular bill is bulkier than the small bill of the former, but never as bulky as that of the latter. Opportunistic and gregarious, often with other finches taking seeds on the ground; also feeds on fruit, buds, food scraps and invertebrates.

Where to see Various habitats on most Galápagos islands, except Española, Genovesa, and remote Darwin and Wolf.

Common Cactus-finch *Geospiza scandens* 12–14cm

This species has the longest bill of all finches, and it is also pointed and slightly drooped. Males mostly blackish. Females heavily streaked and scaled, with a faint white crescent above eyes. Strongly associated with *Opuntia* cacti; feeds on flowers, seeds, nectar and fruit of the cacti, but sometimes descends to ground to pick invertebrate prey and seeds.

Where to see Arid lowlands on several Galápagos islands, including Floreana, Isabela and Santa Cruz.

Buff-throated Saltator *Saltator maximus* 20–21cm

Saltators are largish passerines, midway between a tanager and a cardinal. Buff-throated is the commonest and most widespread. Its bright olive upperparts, black moustachial and buff throat are distinctive, but the relatively common Blue-grey Saltator of Amazonia is superficially similar. A proficient singer, it regularly joins mixed-species flocks but also visits artificial fruit feeders.

Where to see Humid forests, borders and adjacent clearings in Amazonian and Pacific lowlands and foothills.

Blue-grey Saltator *Saltator coerulescens* 20–21cm

Like Buff-throated Saltator, Blue-grey is a proficient singer. Loud mellow voice that is easily imitated. Upperparts ash grey, eyebrow white and short. Mostly in pairs, with or apart from mixed-species flocks. Less wary than other saltators, it often perches on exposed branches, especially to sing.

Where to see Forest borders, woodland, shrubby clearings and cultivated fields in Amazonian lowlands and foothills.

Streaked Saltator *Saltator striatipectus* 19–20cm

An attractive saltator of dry forest, woodland, shrubby clearings and gardens. Birds in south-west are plain olive above, whitish below, and have a long white eyebrow. Birds in Andes have greyish-olive upperparts, densely streaked underparts and a short narrow eyebrow. Juveniles are all streaky, irrespective of region. Regularly in pairs, feeding mostly on fruit and petals. Voice loud and musical.

Where to see Inhabits dry valleys north of Quito, also dry forest and scrub near Guayaquil, Portoviejo and other lowland cities.

Andes

west